D0585992

ENVIRONMENTAL ASSESSMENT

Environmental Assessment

A Practical Guide

C. A. Fortlage

Gower Technical

Published by
Gower Publishing Company Limited
Gower House
Croft Road
Aldershot
Hants GU11 3HR
England

Gower Publishing Company
Old Post Road
Brookfield
Vermont 05036
USA

British Library Cataloguing in Publication Data
Fortlage, C.A.
Environmental assessment: a practical guide.
1. Great Britain. Environment. Effects of man.
Assessment
I. Title
304.2'8'0941

Library of Congress Cataloging-in-Publication Data
Fortlage, C.A.
Environmental assessment : a practical guide / C. A. Fortlage.
p. cm.
1. Environmental impact analysis. I. Title.
TD194.6.F67 1990
333.7' 14—dc20   CIP 89–23749

ISBN 0 566 09045 7

Printed in Great Britain by
Billing & Sons Ltd, Worcester

Contents

CONTENTS

Figures

Acknowledgements

I should like to thank the following people for their kind advice and assistance in the preparation of this Guide: Professor Derek Lovejoy for helpful discussions on the meaning of Environmental Assessment and the role of the landscape designer; Elizabeth Phillips of Architects Information Services for information and advice on planning and environmental legislation; Derek Lovejoy and Partners for the opportunity to study examples of Environmental Assessment in practice, and especially Barbara Brewer for helpful comments during the writing of this book. I should also like to thank Ann Beer, Head of the Department of Landscape Architecture at Sheffield University for advice on the needs of students in Environmental Assessment.

Although there is a considerable literature on Environmental Assessment in the United States, there is little practical guidance for the UK practitioner, and I am indebted to the following publications for information and guidance, although I have not found it appropriate to include any extracts from them in this book.

The most important is undoubtedly:

A Manual for the Assessment of Major Development Proposals, (HMSO, London 1981). Although this was written before the introduction of Environmental Assessment legislation, it provides comprehensive guidance on all aspects of environmental surveys, checklists, techniques, and sources of environmental information.

Department of the Environment, *Planning Policy Guidance Notes*, (HMSO, London, various dates). These embody the Department of the Environment's policies on planning matters. They are not specifically directed towards Environmental Assessment, but they are helpful guides to the official attitudes to development.

Phillips, E.T. and Sarjeantson, M., *Legal Reminders for Architects*, (Butterworth, Sevenoaks 1988). A useful guide to the multiplicity of environmental legislation which may be encountered in an Environmental Assessment.

Standing Advisory Committee on Trunk Road Assessment, *Trunk Road*

ACKNOWLEDGEMENTS

Proposals: A Comprehensive Framework for Appraisal, (HMSO, London 1980). The official Department of Transport method of assessing major road proposals. Thorough and carefully worked out, but with more emphasis on the engineering and traffic aspects than on the natural environment and socio-economic effects. A mandatory system of assessment for all major DoT projects.

Land Use Consultants, *Channel Fixed Link: Environmental Appraisals of Alternative Proposals*, (HMSO, London 1986). An interesting study of the four proposals submitted by contenders for the Channel Tunnel development. It summarizes the Environmental Statements prepared by the consortia, and comments on the quality of the information provided; a valuable lesson in the problems of practical Environmental Assessment.

Research and training in Environmental Assessment is carried out by the Centre for Environmental Management and Planning (CEMP) at the University of Aberdeen. The Centre runs training courses and conferences and produces training manuals for international agencies.

Introduction

This book is intended to provide a basic guide for architects, planners, landscapers and other practitioners in the environmental professions who have as yet had little or no experience of Environmental Assessment procedure and practice. Although the number of major developments in the UK for which Environmental Assessments are mandatory or desirable is likely to be limited, just one project the size of the Channel Tunnel may involve hundreds of people in government departments and local planning authorities, as well as environmental bodies, community groups, and trade unions whether they are opposing or supporting the development. For them and for the thousands of individual landowners, householders, farmers, and shopkeepers who will be affected by development, an understanding of basic principles and current legislation is important.

Environmental Assessment is a discipline which has arisen from man's lack of self-discipline. Before man possessed the power to destroy his environment it was unnecessary to protect the planet from his interference, since the slow pace of development in agriculture and building, and the mastering of water and wind power brought about changes in the environment rather than destruction, allowing time for natural and man-made ecosystems to adapt to the changes. Since the introduction of industrial processes, and the creation of unnatural chemicals and compounds, the environment has been unable to accommodate these rapid and drastic changes; nevertheless, the ability to absorb change is essential in order to maintain the robustness and natural progression of well ordered ecosystems. The assumption in early industry and intensive agriculture – that Nature would always cope, and that even if she didn't, who cared – is no longer acceptable. No voluntary code will control the individuals or nations who are determined to exploit natural resources for their own benefit, and therefore legislation is required to enforce at least the minimum of respect for the environment. To acheive this aim, means must be found for foreseeing, controlling, and mitigating the effects of new forms of interference with the environment. In North

INTRODUCTION

America and Europe, this control is exercised by means of Environmental Assessments of all new major developments which are likely to cause 'significant effects on the environment', a phrase which is now taken to include such purely human concerns as employment, recreation, shopping and education.

Chapter 1

ENVIRONMENTAL ASSESSMENT PAST AND PRESENT

History and development

Environmental Assessment has always been part of any development process – though not under that name, nor in that form – and it is interesting to look at earlier reports and to see how closely they are related to the present-day practice of environmental study and analysis. The historical assessments are different from modern practice, more because of their simple approach and positive conclusions than because of any great difference in philosophy. The demand for some sort of Environmental Assessment has naturally been made by the sufferers from development rather than by the developers, and among the more publicized contemporary resistances to development are the objections made by the people of south-east England to the Channel Tunnel and the high-speed rail link to London. They may seem unusually well organized, but then they have been there before.

In 1548 a Commission was set up to examine the environmental impacts of the Wealden iron mills and furnaces in Kent and Sussex; this body consisted of four chief assessors and 16 other investigators, and they were required to examine the effect of the proliferating iron mills on the economy of Sussex. Most of their evidence was collected from representatives of the towns and districts, probably much like trade unions and amenity groups today, and the factors they considered were very similar to those that would be studied in a contemporary Environmental Assessment: the number of mills; how much wood they consumed yearly; how much the cost of wood had increased due to shortages; which towns would suffer economically from further development; what trades would suffer from timber shortage; how many jobs

1

would be lost; and why the price of iron was higher than it was when there were fewer mills. Their findings were also much like those of today's Environmental Assessment: each mill used about 1,500 loads of wood yearly and made no effort to renew the woodland, so that this resource would soon be exhausted; the trades depending on timber for their raw material were in distress (and in the sixteenth century nearly all household goods and gear included wood in their construction); the fishermen had insufficient wood to build boats or for fires to dry themselves after fishing; and more importantly there was not enough large timber for repairing the harbours and houses. The commissioners were as wary as modern assessors in forecasting future effects, but they predicted that job losses and community decay would eventually be enormous if no steps were taken to mitigate the effects of further development. Their recommendations included restrictions on tree felling and a reduction in the number of mills; though whether these were actually carried out is not certain. What makes this early Environmental Assessment startlingly different from a modern study is the fact that the assessment was *completed* between 13 November 1548 and 14 January 1549, in contrast to Sizewell power station which held the record for the longest planning inquiry.

Environmental Assessment, as we know it today, started in the USA when the expansion of industry which began in the Second World War continued to increase and intensify, and the practices of intensive farming became more widespread. The products and wastes of chemical, manufacturing, and agricultural industries had not previously been so dangerously toxic nor produced in so large a volume as to create serious environmental problems, and the atmosphere, soil, rivers and oceans had been able to absorb them without irreversible damage, while any local effects remained local and did not attract more than specialized protests from naturalists and ecologists who understood the dangers. The post-war expansion created a far different situation, both in the USA and in Europe. It is estimated that in the last 200 years Europe and the USA have been responsible for more environmental destruction than all the rest of mankind since written records began, with the possible exception of the Flood. Even that environmental disaster left no legacy of lethal chemicals for the children of Ham, Shem and Japhet. All ecosystems, including human communities, have thresholds of tolerance for pollution and disturbance beyond which the system may suffer anything from temporary upsets to complete destruction, and the industrial and agricultural practices of post-war industrialists and farmers began to cause environmental damage which crossed these thresholds.

Wide-scale public interest and concern was aroused by Rachel Carson's book *Silent Spring* published first in the USA and then in the UK in 1963. The book set out to show the American people how their land and lives were affected by the large-scale and indiscriminate spraying of crops with powerful insecticides and herbicides. She succeeded in making the public aware of the ecological consequences of introducing toxic chemicals into the natural food chains, and the dire effects of cumulative dosage with apparently small quantities of agricultural poisons. The book is awesome reading even today; accounts of the destruction of nearly all wildlife over hundreds of acres polluted with pesticides are still valuable reminders that environmental effects can never be taken for granted. From this beginning arose *public* concern for the environment (biologists and ecologists had long been aware of the dangers) and, eventually, pressure by the public and environmentalists forced state and federal authorities to exert some control over the release of toxic chemicals into the environment. The control was established by the 1969 National Environment Policy Act, which required Environmental Statements to be prepared for federally-funded or -supported projects which were likely to have impacts on the environment. The exercise of these powers varied from state to state and, in many cases, undesirable developments slipped through the net, interstate impacts being especially hard to control. The US Council for Environmental Quality was charged with the task of developing standard procedures for Environmental Statements, and Europe owes much to the USA for showing the way towards a viable form of environmental control and for pointing out the weaknesses inherent in any system.

This control was much eroded under the Reagan administration, and many environmentally undesirable projects were permitted on economic grounds. It seems hard for Europeans to believe that anyone could seriously suggest pouring millions of gallons of sulphuric acid on to the ground in order to draw off a leachate containing commercial quantities of copper, yet this has been considered to be a viable process in the USA, regardless of the long-term effects on ground water or ecology.

Environmental Assessments can vary enormously in size and scale, depending on the sensitivity of the area and the degree of disruption likely to be caused; the range runs from a manageable disturbance to a village caused by road realignment to the total destruction of irreplaceable fragile ecosystems in the Antarctic caused

3

by the exploitation of minerals. In developed countries there is usually a strong body of informed opinion to counteract any attempt by a developer to slide out from under Environmental Assessment, but in the Third World, and especially in the Antarctic (where the developer is also the national government in charge), the assessment and its findings may be rather less impartial. We have yet to see the introduction of intercontinental or global Environmental Assessments, but, with the increasing level of far-reaching environmental hazards such as ocean dumping of toxic and nuclear wastes, acid rain, nuclear fall-out, ozone deterioration, rain-forest destruction, overfishing, desertification, and other forms of global damage, it is very possible that Environmental Assessments of major developments will be conducted on a scale well beyond our current experience.

Environmental Assessment may be considered as having been accepted in principle at the United Nations Conference on the Human Environment of 1972 at Stockholm when the framework of modern environmental international and national policies was laid down. The Conference generated a concern for the environment which resulted in the 1980 publication of a World Conservation Strategy by IUCN, UNEP, and WWF, and the subsequent launching of a series of national policies on environmental conservation and control, and the slow, controversial but definite progress of the EEC towards formal European legislation. The EEC initiated environmental action programmes in 1973, 1977, and 1983: the principle behind them is that 'the best environmental policy consists in preventing the creation of pollution or nuisances at source', leading to the need to consider developments *before* construction, and the consequent need to create legislation to enforce such consideration. The UK exposition of environmental policy may be found in the set of studies *Conservation and Development Programme for the United Kingdom*, 1983. While many environmental issues such as the deterioration of water quality, the depletion of fish stocks, and migrant bird trapping, are not subjected to Environmental Assessment unless development takes place, the concern of the general public and (under pressure) the government for environmental issues has been stimulated by the publicity given to these problems. In some cases the existence of such problems has a direct influence on the proposals for a project, especially where the scheme affects scarce natural resources or endangered habitats.

The first intimation of hard legislation in Europe on Environmental Assessment came with the EEC Directive 85/337 which formed the basis for most European national laws on environmental control over new developments. Each member country is allowed to integrate the requirements of the Directive with its own planning control system, but any major departure from the schedules of assessable development or the range of subjects to be covered must be submitted to the Council of the European Communities for ratification. The need for legislation was far more apparent in Europe than in the UK, mainly because of the closer geographical relationship which European countries have with one another and the consequent effect which they have on each other's environment. This is particularly evident in dealing with environmental impacts such as pollution of the Rhine, the Danube, the Mediterranean, and the Baltic, where heavy chemical industries can discharge waste products which may easily damage their neighbour's environment. Examples are the discharge of dieldrin to the Baltic from timber works, and the increasing discharge of sewage and industrial effluent into the Mediterranean; so critical have these impacts become that there are projects in hand for cleaning up the Rhine and the Mediterranean, though how far international cooperation can be made to work is not yet clear. The Danube in particular is suffering from the effects of rapid industrial expansion in the Eastern Bloc, where unpurified effluents are being discharged into the river in very large amounts, and huge dam projects[1] are changing the ecology of the river and its margins, thus affecting countries beyond their own borders. In Western Europe the EEC has, however, firmly established the principle that 'the polluter pays' – a policy which is naturally unpopular with those countries like Britain who contribute pollution to the air and the sea without being on the receiving end of any other country's pollutants. Britain is, however, the subject of a large number of claims from European countries who have allegedly suffered environmental damage from UK pollution, and it is likely that the case histories of such claims will become the basis of negotiations between countries who are in dispute over international environmental damage. One difficulty is the long time-scale involved in determining such damage, as effects may not be apparent for decades, and possibly not for generations.

It is clear that, with ever-increasing development of more and larger chemical and processing plants, international litigation over pollution claims would reach an uneconomic level of expenditure, both by private firms and national or regional governments. In

---

1. Unofficial information suggests that this has been stopped. (October 1989)

order to reduce this litigation, and to prevent the occurrence of irremediable environmental damage, the EEC has therefore established the concept and practice of carrying out Environmental Assessments for major projects *before* they are constructed, rather than relying on international law and public outcry to make good environmental damage, particularly as many types of damage are irreversible. The emphasis on Environmental Assessment in the original EEC Directive is accordingly very strongly directed towards pollution and its consequences, with the impacts of projects on employment, social structure, and economics being given rather less consideration, since these factors are not so likely to produce effects across international boundaries. The EEC Directive gives the main priorities for assessment as human health, the quality of life as it is affected by the environment, the continuing diversity of species, and the maintenance of the whole ecosystem. It is noticeable that these priorities seem to be biased towards the continuance and prosperity of the human race, but since we are still not independent of natural resources the other species of animals and plants should benefit almost equally from Environmental Assessment.

In the UK the problem is not so much that of international impacts, since we tend to hope that no one in Europe will be able to identify the source of any pollution they may suffer, but that of satisfying the voting public and the environmental watchdog organizations that projects are environmentally safe. Some of the potential impacts of a new development are obvious to the lay public – visual intrusion, traffic congestion, severance of land holdings, loss of agricultural land, noise, and changes in property values – whilst others require the services of specialists to measure and assess them – hidden pollution, job losses, economic gains or losses, ecological impacts, and effects on scarce natural resources.

Quite apart from the Environmental Assessment regulations introduced by the Department of the Environment, there are other organizations who regulate environmental matters such as the Nuclear Industry Radioactive Wastes Executive (NIREX) and HM Inspectorate of Pollution (which replaced part of the old Factory Inspectorate), and any Environmental Assessment must satisfy the statutory requirements of these authorities as well as meeting the environmental standards of the local planning authority.

Where the public is aware of possible impacts, the level of opposition to the project depends on the current state of public opinion rather than on a long-term and unbiased appreciation of the environment. In an economy where there is a significant level of unemployment, the effect of a project is usually to increase the number of jobs available and to generate 'spin-off' in the form of service and supportive industries, and therefore there is unlikely to be any major opposition from trade unions or the local population. The impacts which usually attract opposition are the impact on the visual quality of the landscape, the pollution and disturbance of the ecology of the area, the land take of houses and agricultural land, and the effect of new infrastructure on human activities.

There have been some attempts in the USA to put a financial tag on each factor in the environment so that environmental impacts can be calculated on a common basis and compared with one another, a method which would also provide a figure which can be used to assess compensation for environmental damage. This has led to such interesting formulae as the rating of a boat at 10 dollars per hour when used for fishing, 5 dollars per hour when used for recreation, and one dollar per hour when moored. Agreement on the value of such activities seems to be almost impossible, and the values would need continual updating, together with seasonal and regional variations to be of any practical use in Environmental Assessment. It is therefore more realistic to deal with each impact in its proper terms, whether these are financial, ecological, social, visual, or political.

The detailed discussion of impacts will be found later in this book, but it should be remembered that each project produces a different range of impacts, both in size, kind, and importance, and consequently a good deal of the emphasis on any one impact depends on current local and political sensitivity. A project which involved the destruction of peat moorland, but provided jobs in return, would meet with approval in northern Scotland, where they have few jobs and much peat, but with strong objections in parts of southern England where full employment and a strong partiality for open landscape are prevalent. Most objections to environmental impacts come from organized groups of people – such as Friends of the Earth, Greenpeace, various conservation societies, and local amenity groups – and it is difficult to know how far they truly represent local or national public opinion. There is thus an obligation to confirm that ideological influences are not distorting the real feeling of the local community.

Meaning of Environmental Assessment terms

The vocabulary of Environmental Assessment is still in a state of development and consolidation; the current meaning of the most usual terms is given here, but like most new disciplines, its practitioners tend to give their own meanings to words and phrases. It is as well to check that everyone on an Environmental Assessment team arrives at a common vocabulary before preparing reports and submissions.

Environmental Assessment

This is the name given to the whole process of gathering information about a project, its possible and probable effects, and the analysis of the data obtained from all sources. As the word implies, it covers much more than the straightforward (or more usually complicated) collection and filing of as much information as can be obtained on all factors relevant to the project. For projects coming under ordinary Town and Country Planning procedures, the developer is responsible for preparing the whole Environmental Assessment, and he is the employer of the Environmental Assessment team. Other Environmental Assessment regulations make the relevant Secretary of State or the developing body responsible for preparing the assessment.

There is no absolute definition of Environmental Assessment as a single concept; it is a compound term embodying ideas and techniques which have developed over many years of increasing concern with the dire consequences of man's interference with the environment. Even the European Communities' legislators have not been able to find a single definition, and descriptions of Environmental Assessment can only be made by listing all the factors which must be considered in connection with a proposed development. Originally the word 'environs' meant the surroundings or neighbourhood of a place; now the word 'environment' has come to be used more in the laboratory sense of the complete set of conditions in which an organism exists, and can cover any situation from the Antarctic perforation of the ozone layer to the local hedgehog's hibernaculum. Up to the middle of this century, the terms commonly used for the environment referred to single aspects of the planet; expressions such as 'Nature', 'the oceans', 'landscape', 'the atmosphere', 'human settlements' are typical of earlier writings. As a wider consciousness of the incredibly complex interactions of all natural and man-made factors has developed, so the use of a single comprehensive term has emerged. 'Environment' is neither elegant nor very explicit, but it has been accepted internationally as being the set of factors in any given situation, their interactions with each other, and with factors outside the situation. Originally the factors were purely natural ones, but human factors such as 'cultural heritage' are now included in the term. It is encouraging to think that the human race (or at least quite a lot of it) has realized that man does not exist as an isolated predator, but is absolutely dependent on his relationships with the rest of the planet, and even with space.

'Assessment' is a simpler term to define. Perhaps humankind is slightly less arrogant than it was; the assessing of a situation implies that not everything is known and not every effect can be predicted with absolute accuracy – a more sensitive attitude than the Victorian conviction, still occasionally put forward, that everything in the universe is capable of rational and exact analysis and definition.

Environmental effect

The original US term for the effect of a development on the environment was 'environmental impact', and this is still much used in the literature of the subject, but Her Majesty's government prefers to use the term 'environmental effect', as this presumably conveys a less drastic impression of the results of development. 'Effects' (or impacts) could be classified in several ways – for example, by their magnitude, their severity, by their beneficial or maleficial effect, or even by their time-scale – but the present UK legislation and circulars do not give any guidance on possible classification, except to distinguish between 'significant' and 'non-significant' effects.

Significant effects

These are effects which disturb or alter the existing environment to a measurable degree. The question as to what is, or is not, a significant effect is one of the most difficult areas of Environmental Assessment to define. The significance of an effect depends very much on the opinion of the assessment team, the local planning authority, the environmental bodies consulted, and the prevailing public sentiment. The best that can be expected is to reach a consensus of opinion amongst those involved in the Environmental Assessment which will reflect current thinking on the importance of environmental factors, but the Regulations imply that the final selection of significant effects lies with the local planning authority (or the Secretary of State).

9

Adverse effects

These comprise those effects which are expected to cause destruction or deterioration of those sectors of the environment which will be affected by the development, either directly or indirectly. Adverse effects are often divided into temporary effects, usually caused by the exploration or construction process, and permanent adverse effects due to the processes used in the development or even merely to its existence. Some adverse effects are irreversible, as occurs in the destruction of a woodland ecology, and some may be reversed to a certain degree over a period of time as the local ecosystem adjusts to the development.

Mitigation of adverse effects

This term refers to the methods proposed by the developer for reducing, obviating or otherwise ameliorating the effects of the project which would have some undesirable effect on the environment. Mitigations are related to one or more of the effects described in the Environmental Statement, and may include technical processes designed to reduce the emission of toxic gases, screen planting to conceal intrusive buildings, regulating the amount of natural resources used, replacing lost jobs with other types of employment, or providing some compensation to the local community for the loss of an amenity. It is arguable how far the mitigation proposals should go in reducing adverse effects – some authorities hold that mitigation should be total, while other experts are willing to settle for a reasonably practical level of mitigation – and, in practice, the level required of the developer will probably depend on the policies of the local planning authority.

Environmental information

This is defined in the principal Regulations as including both the Environmental Statement and any other representations about the likely environmental effects of the project made by interested parties. More broadly, the term comprises all the data on the project which form the raw material for the Environmental Assessment. There is no legal requirement or Department of the Environment (DOE) restriction on the amount, kind, or source of information which may be gathered on a project, and there is no set format for the way in which it is to be presented, but there are obviously some difficulties in collecting certain types of information. No sane industrialist is going to reveal details of his manufacturing process which would help his rivals in business, nor is any conservation group going to admit that they themselves never make use of the amenity they are so ardently protecting. This is not to say that the truth about these factors cannot be discovered, or that they are therefore not relevant, but certainly a Sherlock Holmes talent for observation, exploration, and deduction is a great asset to the Environmental Assessor. Although the assessment team is employed by the developer, they naturally have a professional responsibility to collect and present data impartially, and a friendly informal approach to individuals and organizations which stresses this impartiality will often produce helpful information where a more rigid approach may fail.

One responsibility of the Environmental Assessment team is that of classifying the information they obtain as 'hard' (provable data from irreproachable sources which can be cross-checked), 'intermediate' (information which appears to be sound and reliable but is not capable of exact proof), and 'soft' (information containing subjective opinions and estimates which may be very important in assessing effects, but cannot be proved one way or the other). Very often, 'soft' information comprises the most strongly-felt estimates of environmental effects such as visual intrusion, the destruction of landscape ambience, or the disruption of community spirit.

Much, if not most, of the critical information about the potential effects of a development lies in the hands of official and semi-official bodies such as local planning authorities, the Countryside Commission, the Nature Conservancy Council, the Historic Buildings and Monuments Commission, the Forestry Commission, and similar quangos. There is provision in most Environmental Assessment regulations for obtaining information from these organizations although, as it must be paid for, it may prove expensive. (In passing, it should be mentioned that the operation of collecting Environmental Information, carrying out an Environmental Assessment, and preparing the Environmental Statement, is neither cheap nor rapid, and plenty of contingency time and money need to be included in the budget by the developer.)

To the inexperienced Environmental Assessor, the ownership of an impressive collection of multi-megabyte disks, filing cabinets full of print-outs, and an elaborate data base and spreadsheet permitting every possible combination of statistical analysis, would seem to be the desired objective for the preparation of an Environmental Statement. Not so. The skill of the specialist in Environmental Information collection lies in knowing exactly what data is relevant, and, as far as is scientifically feasible, just how each estimated effect is likely to affect the environment. Once the main data has been collected and examined, it is very often the case that only a small number of effects appear to be really decisive in determining whether or not the project should be given planning permission; it is these significant effects which should be the subject of closer and deeper analysis by the Environmental Assessment team, rather than a wider and shallower coverage of *all* the projected effects. It must be remembered, though, that each environmental guardian group, political lobby, amenity association, or development group, will have its own priorities, and each will be aggrieved if these are not thoroughly considered in the Environmental Assessment, so that not even minor effects can be omitted from the study, even if they do not need to be subjected to the fullest possible analysis.

Environmental Statement

This is defined by the principal Regulations as the official document, containing all the information required by Schedule 3 to the Regulations, which is submitted with the planning permission application. This document contains the results of the Environmental Assessment, and the conclusions drawn from it. It forms part of the planning submission documents and cannot legally be treated as a separate consideration. It must be capable of withstanding challenges and could be treated as part of the proof of evidence in a planning appeal or inquiry. Since the Environmental Statement is accessible to the public through the planning register and to objectors in a planning appeal, the main conclusions should be intelligible to lay readers as well as to the expert witness. The Inspector in a normal planning inquiry is competent to judge the value of technical evidence, but in cases where major developments are subjected to the Private Bill procedure rather than to the normal planning process, the Parliamentary Committee is unlikely to consist of experts, though the Members forming the committee are usually knowledgeable in the environmental field. Both the EEC directive and the UK Regulations therefore call for a non-technical summary of the Environmental Assessment to form part of the Environmental Statement; this is an important part of the submission and should be carefully prepared.

Chapter 2

LEGISLATION

The UK Regulations

Under the European Communities Act 1972, the UK is bound to accept EEC Directive 85/337 as the controlling document which lays down the rules for Environmental Assessment of major development in the UK. The application of these rules is consequently effected by Statutory Instrument, and not, as might be expected, by Act of Parliament. This means that the rules governing the scope and practice of Environmental Assessment are laid down by the Civil Service under the overall conditions of the Directive, and are not debated in either House of Parliament, as they would be if the provisions of the Directive were incorporated in a General Act. When the Directive was first formulated, UK legislators argued that the UK planning system already took into account all the environmental factors listed in the Directive, and that there was no need to adopt further controls. The UK was overruled on this matter, but nevertheless firmly placed the main Environmental Assessment procedures within the Town and Country Planning structure, although the Directive has been implemented by means of Regulations made under the European Communities Act 1972 and not, as might be expected, jointly with an amended Town and Country Planning Act.

The result of this formulation is that an Environmental Statement (the formal conclusions of an Environmental Assessment) forms part of the normal application for planning permission, and cannot be considered separately. Only those projects which would already have been subject to development control are covered by the principal Town and Country Planning (Assessment of Environmental Effects) Regulations; other projects such as Harbours and

Highways are dealt with by separate and somewhat different Regulations. Scotland has separate legislation. Although what are known as 'Schedule 1' projects are required to be the subject of Environmental Assessment, 'Schedule 2' projects are assessed at the discretion of the local planning authority (or the Secretary of State) and thus, broadly speaking, any development removed from planning control becomes exempt from Environmental Assessment control, unless it falls under one of the more specific Environmental Assessment regulations dealing with Harbours, Highways or other special projects. There seems to be a certain lack of correlation between the requirements of the EEC Directive and the Environmental Assessment Regulations, since the Regulations and the 1988 General Development Order both permit the Secretary of State to exempt particular projects from Schedule 1, whereas the Directive only permits this if the European Communities Commission and the public are notified of the exemption and the reasons for it. There may also be cases where the General Development Order schedules development as 'permitted' when, by reason of its location or some other factor, it would possibly be subject to an Environmental Assessment; this anomaly has not yet been resolved, but no doubt subsequent legislation will clear up this point. The several different regulations dealing with various types of Environmental Assessment are not consistent; some have their own schedules of development liable to Environmental Assessment, and some refer directly to the EEC Directive; some require the developer to make an assessment, and some allow the developing authority to carry out and then judge the assessment; and some allow the Minister in charge considerable powers, and some restrict his authority. There is apparently no procedure for appeals against the Minister's decision; whether the High Court will be able to consider appeals is not clear. Neither does there appear to be any machinery for any body other than the local planning authority or the relevant Secretary of State or Minister to demand an Environmental Statement; this means that concerned organizations, such as the Nature Conservancy Council or the Countryside Commission, are restricted to the normal planning procedures for making objections. Only the Harbour Works regulations provide for the appointment of an inspector specifically for Environmental Assessment inquiries, and there is no indication that he should possess any special qualifications.

Development corporations and other quangos who may have been given planning powers and development powers are in the position of being able to design a project, acquire the land, prepare an Environmental Statement, and determine their own application unless the Secretary of State desires to intervene.

It is inevitable that these Regulations will be subject to revision as experience develops, and case law builds up. Also inevitable is that new Regulations will be made to cover other classes of development, as more sophisticated research techniques are developed to provide increasingly accurate assessment of the long-term effects of certain types of development on the environment.

Summary of current English Environmental Assessment legislation

Town and Country Planning (Assessment of Environmental Effects) Regulations 1988. SI 1988/ 1199

Made under European Communities Act 1972.
Operational from 1 July 1988.
Applies to England and Wales only.
These Regulations implement EEC Directive 85/337. Environmental Assessment is required for large-scale or environmentally sensitive projects which require planning permission under the Town and Country Planning Act 1971. Projects coming under other Acts are the subject of separate legislation. The Regulations set out the procedures to be followed, the appeals procedures, and contain three Schedules. Schedule 1 lists projects such as large-scale engineering works for which an Environmental Assessment is compulsory. Schedule 2 lists other projects including extractive industries and developments which *may* have a significant environmental effect and, in such cases, the local planning authority determines the need for an Environmental Assessment, with provision for referring disputed cases to the Secretary of State (Environment). Schedule 3 sets out the type and scope of the information which must be provided in an Environmental Statement.

Note These Regulations are discussed in greater detail later in this Chapter

Town and Country Planning General Regulations 1976 SI 1976/ 1419

Local authority development is subject to the same Environmental

15

Assessment control, but it is implemented under these Regulations. The procedure of deciding whether or not a development falls into Schedule 1 or Schedule 2, referring to the Secretary of State for a decision; carrying out statutory consultations, complying with publicity requirements, and the preparation and consideration of the Environmental Statement must all be followed as though the local planning authority were a private applicant.

Development by the Crown is not bound by these regulations, but the Crown will, as a principle, submit a normal Environmental Assessment to the local planning authority at the early consultation stage.

Town and Country Planning General Development Order 1988 SI 1988/ 1813

Made under Town and Country Planning Act 1971 c.78 and Local Government Act 1972 c.70.
Operational from 5 December 1988.
Applies to England and Wales.
This Order gives the Secretary of State powers to vary the conditions of the Town and Country Planning (Assessment of Environmental Effects) Regulations 1988 so that:

- a particular development coming under Schedule 1 or Schedule 2 of those regulations may be exempted from the need to undergo an Environmental Assessment;
- he may arbitrate as to whether or not a proposed development should come under those Regulations;
- he may direct that a particular development should come under those Regulations even if it does not appear to be in Schedule 1 or Schedule 2.

The local planning authority must follow his direction on these matters when considering a planning application.

Note Although these powers are also provided in the principal regulations, this reinforcement serves to draw attention to the fact that the Town and Country Planning Environmental Assessment regulations are still part of the planning process and cannot be treated separately.

Electricity and Pipe-line Works (Assessment of Environmental Effects) Regulations 1988. SI 1989/ 167

Made under European Communities Act 1972
Operational from 9 February 1989
Applies to England and Wales for electricity works; to Great Britain for pipeline works.
These regulations apply to the following:

- the construction or extension of a generating station on any land under s.2 of the Electric Lighting Act 1909, c.34;
- the placing of an electric line other than a service line above ground under s.10(b) of the Schedule to the Electric Lighting (Clauses) Act 1899 c.19.(included in Electricity Act 1947 c.54);
- the construction of a pipeline for oil or gas under the Pipe-lines Act 1962 c.58;
- the diversion of a pipeline that has been used for oil or gas under the Pipe-lines Act 1962, c.58.

The Regulations will apply if the construction or extension of a non-nuclear generating station is under 300 megawatts. Any applicant may ask the Secretary of State (Energy) to decide whether the project justifies an Environmental Statement, and must submit basic information on the site and the development with his request, and further information if requested. When the Secretary of State considers that the developer has provided enough information, he must consult the local planning authority, who must give their opinion within three weeks. The Secretary of State must give his decision to the developer within three weeks of completing his consultations, giving his reasons for requiring an Environmental Assessment. Should the developer apply for consent to electricity or pipeline works without adding an Environmental Statement, and the Secretary of State thinks that one is necessary, then he must demand it within three weeks. Unless the developer agrees to provide an Environmental Statement within three weeks, his application is deemed to be refused.

The information which must be provided in the Environmental Statement is set out in the Schedule, which is almost identical with Schedule 3 of the Town and Country Planning Environmental Assessment regulations. The Secretary of State may ask the applicant for further information and also for additional evidence to prove his statements. When the statement is completed the developer must

17

publish a notice for two weeks running in two local newspapers, describing the development and stating where and when copies of the statement can be had, and their cost. This is in addition to any other notices required by law.

Note These regulations provide less scope for consultation with environmental organizations than do other regulations. No-one is consulted *after* the Environmental Statement has been circulated; no provision is made for the public to make representations, and no-one except the local planning authority need be consulted at all. Nearly every other statutory instrument requires the local planning authority or the Secretary of State (or the Minister) to consult bodies who have 'environmental responsibilities', and sometimes named organizations. On the other hand, the CEGB has been under an obligation to consider environmental matters since the 1957 Electricity Act, and they have carried out their duties very thoroughly and responsibly.

The schedule gives the statutory contents of the Environmental Statement, which are almost exactly those of the Town and Country Planning Regulations, with the rather odd exception that mitigating measures must be proposed for the 'significant effects', and not just the 'significant adverse effects'. This variation does not seem to imply any great change in the requirements.

Environmental Assessment (Afforestation) Regulations 1988. SI 1988/ 1207

Made under European Communities Act 1972.
Operational from 15 July 1988.
Applies to England and Wales only.
This Regulation deals with the applications for grants or loans in respect of forestry planting. If the Forestry Commission think that such a forestry project is likely to have significant environmental effects, or adverse effects on the ecology, they must take the environmental information into account before making a grant or loan. Any applicant for a grant or loan may ask the Forestry Commission whether or not they think that an Environmental Assessment is required, and must submit a brief description and plans. The Commissioners must give a decision on the need for an Environmental Assessment, and their reasons, within four weeks; if they fail to do so it is presumed that no assessment is necessary. The applicant may appeal against their decision to the Minister (Ministry of Agriculture, Food and Fisheries) who must give a

direction within four weeks. The Minister may also override the Commissioner's decision and insist on an Environmental Assessment.

If an application for grant or loan is made without an Environmental Statement, and the Commissioners think that one is necessary, they must tell the applicant so within four weeks, and the applicant then has another four weeks to appeal against the decision to the Minister.

When the applicant has agreed to prepare an Environmental Statement, he must advertise in two local newspapers, notifying the public that they may make representations on the project within 28 days of the notice; details of the project and the Environmental Statement must also be available to the public for 21 days after the notice is published. The address and the charge for copies of the statements must be given.

When the Commissioners have received the Environmental Statement they must consult the Nature Conservancy Council and the Countryside Commission, or the Countryside Commission for Scotland, as well as any other bodies they think fit, giving them four weeks to make comments. The applicant must be told who these bodies are, and he may ask them for information; in turn they may ask him for more information on the project. The Commissioners may ask the applicant for more information if they think that it is necessary in order to come to a decision. After considering the Environmental Statement and any representations from other bodies, the Commissioners must inform the applicant of their decision and publish it in the same two local newspapers. The Schedule to the Regulations sets out the information to be provided in the Environmental Statement.

Note These regulations deal with grant and loan assisted forestry only; anyone who establishes forestry plantations with his own money appears to be exempt from the need to submit an Environmental Statement, since large-scale forestry is only scheduled in Annex II 1(d) of the EEC Directive and not in the Town and Country Planning Environmental Assessment regulations. This would also seem to imply that the Forestry Commission itself is exempt from the need to carry out an Environmental Assessment, despite the fact that major forest planting is known to have wide repercussions on the environment, ranging from total visual intrusion into the existing landscape to the loss of jobs in sheep management. The Schedule is almost identical to the Town and Country Planning Environmental Assessment Schedule 3, including the need to describe 'the main characteristics of the production processes', which may indicate some control over timber conversion processes. In these regulations, architectural and archaeological heritage are considered as material assets in accordance with the Directive, and not as part of the cultural heritage as in the Town and Country Planning Environmental Assessment Regulations.

Land Drainage Improvement Works (Assessment of Environmental Effects) Regulations 1988. SI 1988/ 1217

Made under European Communities Act 1972.
Operational from 16 July 1988.
Applies to England and Wales only.
Implements EEC Directive 85/337 for land drainage improvement works proposed by drainage bodies, including local authorities acting as drainage authorities. These Regulations apply to all watercourses of every type; rivers, streams, ditches, drains, cuts, culverts, dikes, sluices and sewers, except Public Health sewers. The improvement works definition covers almost any work to a watercourse, including works to control water as well as work to the watercourse itself. The drainage body must consider whether the proposed improvement works are likely to have any significant environmental effects; if so, they must publish a notice giving a description of the project in two local newspapers, stating whether or not they intend to prepare an Environmental Statement. If they do not so intend, it is for the public (or any interested organization) to make representations to the drainage body concerning possible environmental effects within 28 days of the notice. If no representations are made, the drainage body can go ahead with their project without making an Environmental Statement. If, after representations have been made, the drainage body still wants to go ahead without an Environmental Statement, the Minister (Ministry of Agriculture, Food and Fisheries) will give a direction on the matter; he may also ask for additional information.

When the Environmental Statement is complete, the drainage body must announce the fact in two local newspapers, stating when and where the Statement may be seen, and also send copies to the Nature Conservancy Council and the Countryside Commission, as well as to any other bodies they think fit. The public may also have copies on request at a reasonable charge. Any representations on the Statement must be sent to the drainage body within 28 days of publication of the notice.

The Environmental Statement and any representations are then considered by the drainage body and, if all objections are withdrawn, the improvement works can go ahead. If all objections are *not* withdrawn, then the Minister must give a decision; he may ask for further information if required, and may then either give his consent to the improvement works or refuse his consent, giving his reasons for the refusal. The Schedule to the Regulations sets out the matters to be dealt with in the Environmental Statement.

Note There are some notable differences between these Regulations and the Town and Country Planning Environmental Assessment Regulations. The drainage body is not *compelled* to consult any other organization in the preparation of an Environmental Statement, as any consultation is voluntary, neither are they obliged to send copies of the Environmental Statement to any bodies but the Nature Conservancy Council and the Countryside Commission. It seems that the public will have to provide evidence that there are likely to be significant environmental effects from the works, and to do this within 28 days of seeing it in the newspapers. Furthermore, they do not have powers to ask the drainage body for further information to supplement the brief published description. The drainage body itself decides whether or not the environmental effects are sufficiently adverse to warrant cancelling the project and, provided that all objections have been withdrawn, they may go ahead even though there may be adverse effects which have not been represented by objectors. The Town and Country Planning Environmental Assessment Regulations keep a balance between consent and refusal of permission, but the drainage regulations appear to be weighted in favour of consent *unless* objections are put forward. Although the Schedule gives the required contents of the Environmental Statement, this is just a slightly amended copy of the Town and Country Planning Regulations Schedule 3, and is not specifically designed for drainage improvement works. Schedule 2 of those regulations also lists canalization or flood-relief works, and dams or water-holding installations as being liable to Environmental Assessment, and it is not clear which set of regulations is the dominant one. Presumably the Secretary of State for the Environment and the Minister for Agriculture, Food and Fisheries will determine borderline cases.

Environmental Assessment (Salmon Farming in Marine Waters) Regulations 1988. SI 1988/ 1218

Made under European Community Act 1972.
Operational from 15 July 1988.

ENVIRONMENTAL ASSESSMENT

These regulations implement Directive 85/337/EEC, Annex II, 1(g) for the purposes of assessing the effects of salmon farming, but the relevant consenting authority is the Crown Estate Commissioners, not the local planning authority. Marine waters include tidal waters within the territorial limits, but not inland waters or those controlled by the local planning authority. The Crown Commissioners receive applications to farm salmon, and they decide whether an Environmental Assessment is necessary: if so, they must consider the Environmental Statement before reaching a decision. The Commissioners must tell the applicant to prepare an Environmental Assessment within six weeks of his application, and Schedule 2 lists the bodies who must be consulted, and who must also be supplied with information by the applicant. These are:

- any local planning authority adjoining the site;
- the Secretary of State for Scotland and Countryside Commission for Scotland in the case of Scottish waters;
- the Secretary of State for the Environment and the Countryside Commission in the case of English or Welsh waters;
- the Nature Conservancy Council;
- any river purification board adjoining the site;
- any water authority adjoining the site;
- if the site is landward of a line between Burrow Head and St Bees Head then:
  - North West Water Authority,
  - Solway River Purification Board,
  - Secretary of State for Scotland,
  - Countryside Commission for Scotland,
  - Secretary of State for the Environment,
  - Countryside Commission.

The Environmental Statement must contain the information listed in Schedule 1, which is almost identical to Schedule 3 in the Town and Country Planning Environmental Assessment regulations.

When the Environmental Statement is submitted, the Commissioners must publish a notice in the *Gazette* and a local newspaper, stating which Post Office holds a copy of the Environmental Statement, where copies can be had, and the cost. Anybody may make representations within 28 days of the notice. The bodies listed in Schedule 2 must all have copies of the Environmental Statement, and the Commissioners must consult them, giving 28 days for observations. The Commissioners may call for further information and for confirmation of statements, and when they have reached a decision they may attach conditions to the consent. All parties to the Environmental Assessment must be informed of the Commissioners' decision.

Note These regulations include the need for an assessment of the 'Cultural Heritage', which presumably refers to Heritage Coast areas.

Highways (Assessment of Environmental Effects) Regulations 1988. SI 1988/ 1241

Made under the European Communities Act 1972.
Operational from 21 July 1988.
Applies to England and Wales only.
These Regulations amend the Highways Act 1980 c.66 by inserting a new section 105A which requires an Environmental Assessment to be made when certain new highways or major improvements to existing highways are proposed. The Secretary of State must determine whether or not a highway project comes under Annex I or Annex II of the EEC Directive 85/337, and whether an Environmental Assessment is required. If so, he must then publish an Environmental Statement at the same time as the publication of the project details. The Environmental Statement must contain the information set out in Annex III as far as it is relevant, and at least:

- a description of the project;
- the measures which will mitigate any adverse environmental effects;
- the environmental data;
- a non-technical summary of the environmental information.

The public must be given an opportunity to comment on the Environmental Statement. In certain cases the appropriate environmental bodies must be consulted if the project lies in, or within 100m of:

- a National Park or
  Nature Reserve: the Countryside Commission.
- a Conservation Area: the local planning authority.
- an SSSI or other
  protected area under
  s.29 of the Wildlife and
  Countryside Act 1981:   the Nature Conservancy Council.

The Secretary of State must consider the representations made by the environmental bodies and the public before reaching a decision.

Note No timetable is given for the Environmental Statement public consultation process, such as it is. The usual environmental bodies such as the Nature Conservancy Council and the Countryside Commission, do not have to be consulted unless the project lies within a very limited range of protected areas, so that any other consultation process relies on the provisions of the Highways Act and its amendments. It is presumably up to the local planning authority to identify other sensitive areas where highway projects would have significant environmental effects.

Harbour Works (Assessment of Environmental Effects) Regulations 1988. SI 1988/ 1336

Made under European Communities Act 1972.
Operational from 3 August 1988.
Applies to England and Wales only.
These Regulations amend the Harbours Act 1964 c.40 so as to allow the Secretary of State to decide whether or not harbour works are to be dealt with under Environmental Assessment procedures. These regulations refer to EEC Directive 85/337 for the list of Annex I and Annex II projects set out in that Directive. If an application for a 'harbour revision order' is made under the Harbours Act, the Secretary of State (Transport) decides whether the project is Annex I or Annex II. If it comes into a category requiring Environmental Assessment, he may then direct the applicant to provide the information set out in Annex III as far as may be relevant to the project. This must include at least the following: a description of the project; a description of mitigating measures proposed; the data necessary to assess the main environmental effects; and a non-technical summary. The Secretary of State may consult any body with 'environmental responsibilities' he thinks fit, and also give them what information he thinks fit. He must consider all this information when reaching his decision, and also publish that decision. The procedures are otherwise as set out in the Harbours Act.

Note These Regulations are unusual in referring directly to Directive 85/337/EEC rather than including a Schedule of Developments liable to Environmental Assessment.

Harbour Works (Assessment of Environmental Effects) (No. 2) Regulations 1989. SI 1989/ 424

Made under European Communities Act 1972.
Operational from 16 March 1989.
Applies to England, Wales and Scotland.
These Regulations cover harbour works which are below the 'low water mark of medium tides' and which do not come under planning control, the harbour revision order procedure in SI 1988/1336, or an enactment for powers to carry out harbour works. This includes applications to carry out work under s.34 or s.35(1)(g) of the Coast Protection Act 1949 c.74, and under s.37 of the Merchant Shipping Act 1988 c.12. Four different Ministries are involved:

- for fishery harbours in England: (Ministry of Agriculture, Food and Fisheries).
- for fishery harbours in Wales: Secretary of State for Wales.
- for marine works in Scotland: Secretary of State for Scotland.
- for any other harbour works: Secretary of State for Transport.

The appropriate Minister must decide if the harbour works come under these Regulations, and if so, whether the project comes into a category listed in Annex I or Annex II of EEC Directive 85/337/EEC, and whether an Environmental Assessment is required. The Minister may require the developer to provide basic information on the project in order to reach a decision. If the Minister decides that an Environmental Assessment is required, the developer must provide the information scheduled in Annex III to the Directive as far as is relevant to the project. This must include at least the following: a description of the project; a description of mitigating measures proposed; the data necessary to assess the main environmental effects; and a non-technical summary. The developer must publish a notice in a local newspaper in the area of the project, at least 14 days before submitting his Environmental Statement, describing the project and giving times, dates, and places where the statement can be seen and where copies can be bought (and the charge for them). The statement must be available for at least 42 days after the notice, and anyone may make representations to the Minister within seven days after that. The developer must also post site notices for 42 days. The Minister may direct the developer to give information to any bodies with 'environmental responsibilities' he thinks fit, and he must also consult those bodies before making a decision. The Minister may appoint an Inspector to examine the proposal, and the result of the inquiry must be considered in reaching a decision. If harbour works are carried out without seeking a decision on Environmental Assessment, the work may be stopped and the Minister may reinstate the site himself and charge the developer. Conditions may be imposed for a fixed period or indefinitely.

Note These Regulations incorporate some of the usual Town and Country Planning procedures such as posting notices, certifying the posting, appointing Inspectors and enforcing control of illegal operations, since the developments covered by them do not come under planning control. The Regulations are unusual in referring to Directive 85/337/EEC rather than including a Schedule of developments liable to Environmental Assessment. The Town and Country Planning Environmental Assessment Regulations also include harbours, fishing harbours, and yacht marinas in Schedule 2. Presumably the Secretary of State for Transport and the Secretary of State for the Environment will have to determine cases on the borderline.

Summary of current Scottish Environmental Assessment legislation

The Regulations for Scotland are somewhat different to the English ones, since one set of Regulations covers Town and Country Planning, Electricity, Roads and Bridges, Development by Planning Authorities, and Land Drainage, although the principles remain the same. The principal Regulations are set out below.

Environmental Assessment (Scotland) Regulations 1988.
SI 1988/ 1221

Made under European Communities Act 1972.
Operational from 15 July 1988.
Applies to Scotland only.

These Regulations cover planning permission applications made under the Town and Country Planning (Scotland) Act 1972 c.52 and the Town and Country Planning (Development by Planning Authorities) (Scotland) Regulations 1981 (SI 1981/ 829 amended by SI 1984/ 238). The Regulations sometimes refer to EEC Directive 85/337 Annex I and Annex II developments; in Part II Planning, Part III Electricity, and Part IV New Towns these should be read as referring to Schedule 1 and Schedule 2, while in Part V Drainage no Annexes or Schedules are mentioned. In Part VI Roads the original EEC Directive Annexes should be referred to, but copies of these are not included in the document.

The main variations from the English Regulations are as follows:

1 Part II Planning. Where the English Regulations allow three weeks for decisions by the local planning authority or the Secretary of State, or their replies to requests by the applicant; the Scottish Regulations allow four weeks.

They include a clause allowing statutory consultees to withdraw formally from the consultation process if they consider that consultation is not required in a particular case or class of development.

The regional planning authority may call in planning applications which require an Environmental Assessment.

2 Part III Electricity Applications made under various Electricity Acts. These are similar to the English Electricity Regulations, although if, after a request for a direction on the need for an EA, the Secretary of State fails to give one, the development is then deemed to require an EA. In these Regulations, it is the planning authority or the Secretary of State who notifies the statutory consultees that information must be supplied, although the applicant may do so if he wishes.

3 Part IV Specific Developments in New Towns, and procedures for Environmental Assessments in New Towns. Development Corporations will act as planning authorities for the purposes of these regulations, which are very similar to the general town planning regulations in Part II.

4 Part V Drainage works under the Land Drainage (Scotland) Act 1958 c.24. These Regulations cover drainage works which are considered by local authorities or statutory bodies to be undesirable on environmental grounds. The Secretary of State cannot then give an improvement order for drainage works unless an Environmental Assessment has been carried out.

5 Part VI Trunk road projects under the Roads (Scotland) Act 1984 c.54. This Part of the Regulations amends that Act to the effect that the Secretary of State decides whether or not the road proposals come under the Directive. If the roadworks are then deemed to come under Annex I or Annex II, the Secretary of State has to prepare the Environmental Assessment and Environmental Statement, and must allow the public time to give their opinions before the project can start. These opinions and the Statement must be considered before any decision is made. Generally these regulations follow the lines of the English Highways regulations.

The Schedules to the Regulations list projects for which an Environmental Assessment is required, and give a list of the information required. Schedule 1, for which Environmental Assessments are mandatory, is the same as that in the English regulations, with the major addition of nuclear power stations and reactors (except research installations). Schedule 2, for which an Environmental Assessment is only required if the environmental effects are likely to be significant, covers the same developments as that in the English regulations. Schedule 3, with slightly different wording, lists the same Environmental Statement information as that in the English regulations.

Town and Country Planning (Development by Planning Authorities) (Scotland) Regulations 1981 (SI 1981/ 829 amended by SI 1984/ 238)

Local authority development is subject to the same Environmental Assessment control, but it is implemented under these Regulations. The procedure of deciding whether or not a development falls into Schedule 1 or Schedule 2, referring to the Secretary of State for a decision; carrying out statutory consultations, complying with publicity requirements, and the preparation and consideration of the Environmental Statement must all be followed as though the local planning authority were a private applicant.

Town and Country Planning (General Development) (Scotland) Amendment (No. 2) Order 1988. SI 1988/ 1249

Made under Town and Country Planning (Scotland) Act 1972
Operational from 10 August 1988
Applies to Scotland.

This Order enables the Secretary of State to direct that development is exempt from the General Development Orders 1981 to 1988, in the case of development set out in the Environmental Assessment (Scotland) Regulations 1988 Schedules 1 or 2. The Secretary of State may determine whether or not a particular project is to be the subject of an Environmental Assessment before planning permission can be granted, and that an Environmental Statement must be considered before planning permission can be granted.

Note These Regulations are similar in effect to the English General Development Order provisions described above.

Department of the Environment Circulars

In order to supplement the Regulations themselves, the Department of the Environment has produced Circulars expounding the Regulations, and in some cases, putting forward thresholds for Schedule 2 projects above which the DoE considers that an Environmental Assessment is likely to be necessary, either because of the size of the project itself, or because of the scale of the expected environmental effects.

Department of the Environment Circular 1988/ 15. Environmental Assessment. Department of the Environment 1988

This Circular explains the Town and Country Planning (Assessment of Environmental Effects) Regulations 1988 which implement the EEC Directive 85/337. The document discusses the procedure for submitting Environmental Statements with planning applications and their publicity. Generally the local planning authority decides whether or not an Environmental Assessment is required for a Schedule 2 project. The Ministry of Defence will prepare Environmental Assessments relevant to their proposals, and the Ministry of Transport will deal with motorway service areas. Guidance on Environmental Assessments in Enterprise Zones, Simplified Planning Zones and New Towns will follow. All Environmental Assessments must be notified to the Secretary of State by the local planning authority, and the authority must enter them in the planning register.

ENVIRONMENTAL ASSESSMENT

Department of the Environment Circular 1988/ 24, Environmental Assessment of Projects in Simplified Planning Zones and Enterprise Zones. Department of the Environment 1988.

Schedule 1 projects – that is, those in Schedule 1 to the Town and Country Planning (Assessment of Environmental Effects) Regulations 1988 (SI 1988/ 1199) which are proposed for development in Simplified Planning Zones – must be the subject of Environmental Assessment. (Enterprise Zones were established before Environmental Assessment Regulations came into force, and therefore do not come under this legislation.) In Simplified Planning Zones Schedule 2 projects are the subject of alternative procedures: either the local planning authority specifically omits Schedule 2 projects from permitted development; or it requires the developer to state that the project should be exempt from environmental assessment.

The Secretary of State may direct the local planning authority to follow one or other of these procedures. If an Environmental Assessment is deemed to be necessary, a separate planning application will be required and the Environmental Assessment should follow the guidelines set out in DoE circular 1988/15.

Private Bills

There is another procedure for dealing with major projects, not covered in the EEC Directive, and that is the British system of the Private Bill. Public opinion has it that this procedure is a way of avoiding statutory examination of environmentally dubious projects, but in many ways the Private Bill method is superior to the public inquiry on a planning application. Although Private Bill procedure does not call for an Environmental Assessment as such, it is improbable that the environmental issues would be given a less intensive study and assessment than they would get in a statutory Environmental Assessment. The Bill has to pass the scrutiny of a Parliamentary Committee before it is passed by the House – both for the Commons and for the Lords – and the members of this Committee are responsible to the whole House as well as to their constituents for their decisions, whereas the Planning Inspector is only responsible to the Department of the Environment, and may have his decision overturned by the Secretary of State. The Committee proceedings are as much open to the public as a public inquiry, save that the number of people who can attend is limited by the size of the committee rooms in the Houses of Parliament. If there are a large number of witnesses and counsel present, the public may have to be content with being represented by a small number of observers. The Committee members can call for further evidence if they wish, and there is not such a strict schedule for Committee hearings as for a public inquiry, so that any doubtful point can be thoroughly examined. The disadvantage is that there may be a bias for or against the Bill according to the Committee members' party allegiances, but the same objection applies to a Secretary of State for the Environment, since he is a member of the government of the day. The Private Bill, after Committee stage, has to meet the challenge of debate in both Houses of Parliament and to satisfy all MPs that the project is environmentally acceptable. This lengthy process, which may extend over two years or more, makes a Private Bill a very expensive method of approving a development, but it does give plenty of time for both public and press to comment and to propose amendments. The 1988 Joint Committee on Private Bill Procedure has recommended that the Bill procedure should cover development powers only, and that the whole Environmental Assessment procedure should come under the normal Town and Country Planning and Environmental Assessment legislation. The Secretary of State could use his powers to require an Environmental Statement for such developments.

The principal Regulations

Town and Country Planning (Assessment of Environmental Effects) Regulations 1988. SI 1988/ 1199

These are the principal Regulations governing environmental assessment of projects which come under the Town and Country Planning Act 1971, including those proposed by the local authority themselves and the Crown, in England and Wales. The Scottish Regulations are very similar in their effect, although related to the Scottish legal system. The Regulations cover applications for planning permission made under Part II of the Town and Country Planning Act 1971 and the self-approval of applications by local authorities under the Town and Country Planning General Regulations 1976. For the purposes of the Regulations, any authority with planning powers, such as a Development Corporation or Housing Action Trust is a local planning authority.

The key to the requirement for an Environmental Assessment is contained in Schedules 1 and 2: Schedule 1 projects are mandatory candidates for an Environmental Assessment; a Schedule 2 project is only included if it is likely to have 'significant effects on the environment by virtue of factors such as its nature, size or location'. A planning permission application for a Schedule 1 or 2 development which requires an Environmental Statement, cannot be granted unless the local planning authority or the Secretary of State has taken the environmental information 'into consideration'. The fact that a particular development does *not* require an Environmental Assessment does not mean that the local planning authority can omit the consideration of environmental factors; this should be done just as carefully in the case of an ordinary planning application but without the full-scale Environmental Assessment and Environmental Statement procedure. Schedule 2 projects in a Simplified Planning Zone may be covered by an umbrella exemption granted by the local planning authority.

A programme of formal consultation is laid down in the Regulations which ensures that essential information held by government, local government, and quangos (Quasi-national Government Organizations) is made available to the Environmental Assessment team, but there is a strong recommendation, supported by overseas experience, that as much discussion as possible between all parties improves the quality of the Assessment. While it is indubitably the developer's responsibility to produce forecasts of environmental effects, the bodies who supply his information may be more experienced in judging the effects of any development in their field, and their help and advice is invaluable. This is difficult when the information must be obtained from parties who oppose the project altogether, but a sensitive and non-aggressive approach may help. There is one bright spot – no extra planning fee is payable when an Environmental Statement is required.

The sequence of events for a planning permission application with Environmental Assessment would follow approximately the course outlined below.

First, in order to avoid unnecessary, lengthy and expensive studies by developers, the applicant may ask the local planning authority whether they think that the project is Schedule 1 or 2, and, if it corresponds with a type of development listed in Schedule 2, whether they think that its environmental effects are likely to be significant enough to warrant an Environmental Assessment. A summary of the project and its likely effects must be submitted by the applicant with his request for a decision, and the local planning authority may ask for further information, and may consult environmental bodies, but they must give an answer within three weeks of the date of the request. (In most sections of the Regulations, the statutory periods may be extended by agreement between the developer and the local planning authority.)

There is nothing to stop a developer submitting an Environmental Statement without being made to do so, but the information he supplies will be taken into account when the planning application is considered, even if the local planning authority decide that the project is not one which requires an Environmental Assessment. However, if the applicant declares that the document he submits is a formal Environmental Statement, then the local planning authority must accept it as such and treat the development as one which requires an Environmental Assessment.

If, by mistake, or by optimism, or because it is not immediately obvious that the project would be liable to Environmental Assessment, a planning permission application is made without this precautionary request, and the local planning authority subsequently decide that Environmental Assessment is required, they must tell the applicant what Schedule the project falls under, and that they require an Environmental Statement to be submitted with the planning permission application. They may consult environmental bodies, but they must notify the applicant of this decision within three weeks and, if an Environmental Assessment is required, the details of the decision must go on Part I of the Planning Register.

Should the applicant not agree with their decision, or should they fail to give one, it is open to him to appeal within three weeks to the Secretary of State for a ruling, and in that case the normal period for non-determination of permission starts only when an Environmental Statement is submitted, or when the Secretary of State directs that one is not required. If the applicant misses this crucial three week target date, his application is deemed to be refused, and from that refusal there is no appeal.

The period of three weeks which occurs throughout the Environmental Assessment procedures does not appear to have any particular significance. There is provision in the Regulations for extensions of time to be granted by agreement between the applicant and the local planning authority. There seems to be a principle embedded in these Regulations that everybody must have copies of every document; a very careful check of the small print in the Regulations, a large calendar, a comprehensive flow-chart, and a reliable copier are essential for a correct and accurately timetabled application. The planning application may be refused if the applicant fails to submit his Environmental Statement or to notify all the people he should have done by the correct dates.

The local planning authority cannot simply demand an Environmental Assessment; a written reply to the applicant giving their reasons must be made, and this document must be kept with Part 1 of the Planning Register, which is open to the public. Until a formal planning application is made, the requirement for an Environmental Statement is not published in the local press, so it is up to local environmental watchdogs to keep an eye on the Planning Register if they wish to be informed of possible Environmental Assessments. Once a formal application is made, these documents are transferred to the Planning Register.

If the applicant is dissatisfied with the local planning authority's decision, or if they fail to give one within three weeks, the applicant can ask the Secretary of State for a ruling. All the paperwork so far generated between the applicant and the local planning authority must be sent to the Secretary of State, and all further documents must be copied to all parties. The minister may ask for further information in his turn, but he must give *his* decision, with reasons, within three weeks of the demand for a ruling. This decision must be placed with Part 1 of the Planning Register.

If the Secretary of State has an application for planning permission referred to him, or if he calls in a planning application with no Environmental Statement attached, he must tell the applicant within three weeks that one is required, and give his reasons in writing.

If the applicant does not agree to this requirement within three weeks, the application falls by the wayside, and there is no appeal from the decision except through the courts. The same principle applies to planning appeals referred to the Secretary of State and to appeals where the Inspector is unsure whether or not the project is liable to Environmental Assessment and has to get a ruling on the matter.

It is worth while remembering that the Secretary of State has powers to intervene before or after a planning permission application has been made and to require that an Environmental Assessment be carried out; he can also direct that a complete class of development should be subject to Environmental Assessment. This is similar to the 'calling-in' powers used for ordinary planning applications.

Once everyone has agreed that an Environmental Assessment is required before planning permission can be granted, and a summary of the project and its likely environmental effects are in the hands of the local planning authority, a further round of information exchange begins. Many organizations must be consulted in the same way as for a normal planning application, and the local council is not exempt from these consultations even for its own planning applications. The Regulations list the bodies who must be consulted and given all the information on the development already provided by the applicant. The Regulations refer to the 1977 General Development Order; presumably that list is now superseded by the one in the 1988 GDO. A much more comprehensive list is given in *A Manual for the Assessment of Major Development Proposals*.

- Bodies listed in the Environmental Assessment Regulations:
  - the local council (if it is not the local planning authority);
  - the Countryside Commission;
  - the Nature Conservancy Council;
  - Her Majesty's Inspectorate of Pollution (where atmospheric pollution, or radioactive or special wastes are involved).
- Bodies listed in article 18 of the General Development Order 1988:
  - in Greater London or metropolitan county: local planning authority.
  - in or near National Parks: district planning authority.
  - development involving hazardous substances: Health and Safety Executive.
  - development generating extra traffic at trunk road junctions or level crossings: appropriate Secretary of State and, for level crossings, British Rail.
  - development involving material increase of traffic and/or effects on highways: local highway authority.
  - development in coal-working areas: British Coal.
  - mining: water authority concerned.
  - open-cast coal mining: Secretary of State for Energy.
  - development within 3 km of Windsor Castle and Parks, or 800 m of any other royal palace or park: Secretary of State for the Environment.

- listed buildings in Greater London: Historic Buildings and Monuments Commission.

- development affecting a scheduled Ancient Monument: Historic Buildings and Monuments Commission; in Wales the Secretary of State for Wales.

- work in or on the banks of a river or stream: the water authority concerned.

- refining or storing mineral oils: the water authority concerned.

- land used for waste disposal: the water authority concerned.

- sewerage works: the water authority concerned.

- cemeteries: the water authority concerned.

- development in an SSSI: Nature Conservancy Council.

- development on land where there is a theatre: the Theatre Trust.

- non-agricultural development which does not comply with a development plan and results in the loss of not less than 20 ha of grade 1, 2, or 3a land or where development would lead to further loss of agricultural land amounting to 20 ha or more: Minister of Agriculture, Fisheries and Food, or the Secretary of State for Wales.

- development within 250 m of land used for waste disposal within the last 30 years: waste disposal authority.

The Secretary of State may direct that other consultations should be held.

This is not entirely a one-way system. Any of the bodies who have been notified, including the local planning authority, are under an obligation to provide Environmental Information as long as it is not confidential. The degree of confidentiality is not given, but presumably information listed as confidential under the Local Government (Access to Information) Act 1985 c.43 could be used as a basis; confidential information may not be taken into account when the application is being considered. It may be that the Secretary of State is also liable to supply information when he acts as the planning authority, but this is not clearly established. These bodies may make a 'reasonable' charge for their information. The feedback, in the form of comments and opinions, both from these bodies and from the public, must be considered by the local planning authority when deciding the planning application.

Under the Town and Country Planning Act 1971 and its amendments, publicity has to be given to applications for planning permission. These Regulations impose further paperwork (supplementing the normal planning application publicity) on the applicant; if the planning application is accompanied by its Environmental Statement this document must be accessible to the public; an address near the site of the proposed development where copies of the Environmental Statement can be obtained must be given, and the charge for the copies must be stated. The Regulation states that the charge should reflect the cost of production, although no limit is placed on this charge, except that it must be reasonable; however, what is or is not reasonable depends on the point of view. The number of copies to be made available is not stated. (While copying plain text is easy, the duplication of coloured illustrations is expensive. It is not unusual for 100 or more copies to be needed for a major project, and delays caused by printing hold-ups have been known to make more trouble than real snags in the Environmental Assessment itself.)

If, for some reason, the applicant for a Schedule 1 or 2 project has *not* included an Environmental Statement with his application, but is going to submit one later, then a more complex set of publicity exercises must be gone through, including newspaper advertising and site posting. The Environmental Statement must now include confirmation of this advertising and copies of the advertisements. During this period, the three-week limit (now for some reason 21 days) is suspended, and takes effect from the date when the local planning authority receives the documents. The same procedures apply to an appeal.

Whichever route is taken, copies of all documents have now been served on the local planning authority, the statutory consultees, and the Secretary of State, and their names must be given to the local planning authority. (Bearing in mind that the local planning authority or the Secretary of State may ask for further Environmental Information and that new data may appear during the process, it is sensible to record the people who buy Environmental statements off the counter so that they can be sent amendments, although this is not a statutory requirement. Everyone who had the original documents ought to be sent amendments, so it pays to ensure that the Environmental Statement is as complete as possible before distributing it.)

Now the local planning authority takes up the running. They must put the Environmental Statement on Part I of the Planning Register, and check that everyone entitled to a copy has had one. In order to allow for the vagaries of the Post Office (and the photo-copier) the local planning authority cannot determine the application until 14 days after the last recipient has been served with his copy. (It might be good practice to obtain dated receipts from all those who have had copies and the amendments.) The same procedure for disbursing copies of the Environmental Statement applies if the application goes directly to the Secretary of State by way of a determination on Schedule 1 or 2 categories, or by way of an appeal.

If the local planning authority has not given their planning decision within the statutory time limit, the start-date for that time limit is taken as the date of the developer's submission of an Environmental Statement, or the Secretary of State's direction that an Environmental Statement is not required.

Since an Environmental Statement takes a lot longer to check than an ordinary planning permission application, and more people are involved, the local planning authority has 16 weeks instead of eight weeks before they need give a decision, and this period may be extended by agreement. The Environmental Statement and any formal directions or opinions must be placed on Part I of the Planning Register. Although the planning officers will have been involved in the Environmental Assessment from an early stage, the complexity of the factors in an Environmental Assessment means that the members of the planning committee may need more time to study the documents and to form a sound judgement. Some environmental effects are easily grasped, but others take much longer to assimilate, and the planning committee should be given as much professional assistance as possible.

If the local authority itself is making a planning application for a Schedule 1 or 2 project, it is expected to go through much the same procedures as the private applicant, including obtaining a ruling from the Secretary of State if required and carrying out the consultation and publicity exercises described previously. It must not grant itself planning permission for a Schedule 1 or 2 project without taking the environmental information into consideration.

Even when the Environmental Statement seems to have been completed, the local planning authority, the Planning Inspector, or the Secretary of State, may ask for further Environmental Information if they think that it is essential to the proper determination of the application or appeal, *and* if they consider that the applicant should have known this information and should have included it in his Environmental Statement. Any or all of them may ask for further evidence to confirm information given in the Environmental Statement. It is not easy to see how information provided by other bodies can be confirmed, but presumably they can be called upon by the applicant to substantiate their statements.

Finally, an Environmental Statement and the planning application must be sent to the Secretary of State so that he may decide if the development is likely to affect other EEC countries. This is in case the project is going to cause adverse environmental effects in Europe; in this rather unlikely event the planning permission may be held up until the countries affected have been consulted.

There will be an enormous data bank remaining in the Environmental Assessment team's offices and the last really important job is to extract the essential material, put it into order and store it somewhere accessible so that it will be available as a guide for future projects of a similar type. Some money should be set aside in the project budget for this job.

It will be obvious from this brief summary of the Regulations that very careful timetabling and coordination are necessary to ensure that all requirements are satisfied. Last-minute revisions, overdue information, and the 'let's just look at that again' syndrome which frequently afflicts anxious professionals have no place at all in an efficient Environmental Assessment programme.

Chapter 3

SCHEDULES 1, 2 AND 3

Schedule 1 developments

Both the Directive and the subsequent Town and Country Planning Environmental Assessment legislation divide new developments into two main groups: those for which an Environmental Assessment is mandatory, and those for which it is discretionary. The list of developments which *must* be the subject of Environmental Assessment are contained in Annex I of the Directive and Schedule 1 of the Town and Country Planning (Assessment of Environmental Effects) Regulations 1988. There are certain differences between the two lists, and it is interesting to compare them; how far the Council of Europe can enforce compliance with their own list of mandatory developments will become apparent after more projects have been through the Environmental Assessment procedures.

EEC Directive 85/337 Summary of Annex I Projects.	UK Regulations SI 1988/ 1199 Summary of Schedule 1 Projects.
	(1) Building or other operations or material change of use of land or buildings to provide:
1. Crude-oil refineries, installations for conversion of coal or bituminous shale at 500 tonnes per day or over.	1. Crude-oil refineries, installations for conversion of coal or bituminous shale at 500 tonnes per day or over.

EEC Directive 85/337
Summary of Annex I Projects.

2. Thermal power stations or combustion installations of 300 megawatts or over, nuclear power stations, nuclear reactors.

3. Installations solely for the permanent storage or disposal of radioactive waste.

4. Works for initial melting of cast-iron and steel.

5. Installations for extracting and processing asbestos at 20,000 tonnes per year or over, (and other asbestos processes).

6. Integrated chemical installations.

7. Motorways, express roads, lines for long-distance railway traffic, airports with runways 2,100 m long or over.

8. Trading ports, inland waterways and inland waterway ports for vessels over 1,350 tonnes.

9. Waste disposal installations for processing or landfill of toxic wastes.

UK Regulations SI 1988/ 1199
Summary of Schedule 1 Projects.

2. Thermal power stations or combustion installations of 300 megawatts or over. *Not* nuclear installations.

3. Installations solely for the permanent storage or disposal of radioactive waste.

4. Works for initial melting of cast-iron and steel.

5. Installations for extracting and processing asbestos at 20,000 tonnes per year or over, (and other asbestos processes).

6. Integrated chemical installations for olefins, certain acids, chlorine, and fluorine.

7. Special roads, lines for long-distance railway traffic, airports with runways 2,100 m long or over.

8. Trading ports, inland waterways and inland waterway ports for vessels over 1,350 tonnes.

9. Waste disposal installations for incineration or chemical treatment of special wastes.

(2) Landfill for special waste or material change of use for such landfill.

The most important difference between the two sets of projects is the omission in the UK regulations of any nuclear installations. These are dealt with in Britain by the United Kingdom Atomic Energy Authority, and presumably will remain outside the Environmental Assessment regulations.

Schedule 2 developments

The developments in Schedule 2 are roughly similar to those listed in Annex II of the Directive and, since these are discretionary in both cases, there is little point in making a detailed comparison of the two lists; any really interesting differences are noted in the appropriate place in the text.

Schedule 2 developments are far more numerous than those in Schedule 1. An Environmental Assessment is not mandatory for these developments, and is only required if the development is likely to have 'significant' environmental effects. It is for the local planning authority, or in certain cases the Secretary of State, to decide if a Schedule 2 project requires an Environmental Assessment, and it is open to the developer to appeal against their decision. The Schedule is divided into groups of projects by type, but the descriptions are not intended to be absolute and a development may still require an Environmental Assessment even if its description varies slightly from that published. Only the traditional English system of developing case law and the accumulation of precedents will enable more definite schedules to be constructed. In order to help the local planning authority to make this critical decision, the Department of the Environment has published guidelines setting out the circumstances which could affect the scheduling of the development. The groups are as follows.

Synopsis of Schedule 2 of the Regulations	Synopsis of the DoE guidelines in DoE Circular 1988/ 15
1. AGRICULTURE: Water management.	1. AGRICULTURE: If the works involve the drainage authorities (Water Act 1973) (Wildlife and Countryside Act 1981), and affect environmental interests.
Pig-rearing.	Over 400 sows or 5,000 pigs.
Poultry-rearing.	Over 100,000 broilers or 50,000 other poultry.
Salmon-farming (hatcheries and rearing).	Over 100 tonnes of fish per year, if they affect rivers or other environmental effects.
Reclamation of land from the sea.	As part of water management above, and also if flood defences are involved.

Note Annex II also includes:

- use of uncultivated or semi-natural areas for intensive agriculture;
- afforestation liable to cause adverse ecological changes;
- land reclamation for conversion to other land uses.

Synopsis of Schedule 2 of the Regulations	**Synopsis of the DoE guidelines in DoE Circular 1988/ 15**
2. EXTRACTIVE INDUSTRY: Extracting peat, sand, stone, gravel, minerals and quarrying.	2. EXTRACTIVE INDUSTRY: Depends on location, size, duration of extraction, waste disposal, access, nature of plant and ancillary works, working methods, transport of minerals. Operations in an AONB or a National Park should merit an Environmental Assessment.
Deep drilling for thermal energy, water, or for nuclear waste. Oil and gas extraction.	Exploratory drilling may not create environmental effects; the handling of output and location of the site may require an Environmental Assessment. For oil and gas extraction, production over 300 tonnes per day may merit an EA.
Deep coal or mineral mining.	New large deep mines may need an EA.
Open-cast mining.	Coal and sand or gravel workings with sites over 50 ha and smaller ones if in a sensitive location or are obtrusive.
Industrial plant for extraction of coal, oil, gas, petroleum, ores, bituminous shale.	Handling of output and location of the site may require an EA. Extraction over 300 tonnes per day may merit an EA, or where the site is in a sensitive area.
Coke ovens and cement works.	

ENVIRONMENTAL ASSESSMENT

Synopsis of Schedule 2 of the Regulations

3. ENERGY INDUSTRY:
Thermal power station below Schedule 1 level, and installations for producing electricity, hot water, steam. Hydro-electric installations.

Industrial installation for transporting gas, steam, hot water, electricity including overhead cables.

Underground or surface storage of gas or fossil fuels.

Making coal or lignite briquettes.

Installation for production or processing of nuclear fuels, and radioactive waste not included in Schedule 1.

4. MANUFACTURING INDUSTRY:
Steelworks below Schedule 1 level, other large metalworks such as smelters, foundries, or rolling mills, boiler-making, car and air-craft, shipyards, railway shops, explosives swaging, roasting of metallic ores.

Glass manufacture.

Production and processing of chemicals below Schedule 1 level, and pesticides, pharmaceuticals, paints, varnishes,

Synopsis of the DoE guidelines in DoE Circular 1988/ 15

3. ENERGY INDUSTRY:
No specific thresholds or criteria are set for developments in the energy industry category.

(There are separate regulations for transmission lines and pipelines.)

4. MANUFACTURING INDUSTRY:
Sites over 20 ha may need an EA. An EA may also be required:
• if the process is a 'scheduled process' as far as air pollution control is concerned;
• if the process discharges any wastes to water, which need water authority consent;
• if the development generates significant amounts of hazardous polluting substances;
• if the process generates radioactive or hazardous waste;

**Synopsis of Schedule 2
of the Regulations**

**Synopsis of the DoE guidelines
in DoE Circular 1988/ 15**

elastomers, storage of chemical, petrochemical products.

- if the location, type, and
extent of the emissions or
wastes from manufacturing
plant may have a signifi-
cant effect on the environment;

Her Majesty's Inspectorate of
Pollution, the Health and
Safety Executive, the water
authority, and the environmental health authority may give
useful advice.

Manufacture and processing
of most food products, including breweries, slaughter-
houses, fishmeal and sugar
factories, and dairy products.

The pollution controls exercised by other legislation are
quite independent of Environmental Assessment; the
requirement to comply with
these controls does not imply
that an EA is needful, or
conversely, that no control
means no Environmental
Assessment.

Processing raw wool, manufacture of paper and board,
dyeing works, cellulose
production, tanneries and
leather works.

Rubber Industry.

5. INFRASTRUCTURE PROJECTS:
Industrial estates, urban developments, ski or cable lifts,
roads and harbours and aero-
dromes below Sch. 1 level,
dams, canals and flood relief
works, surface or under-
ground passenger tram or railway below Sch. 1 level, oil

5. INFRASTRUCTURE PROJECTS:
(a) Industrial Estates
Industrial estates may incur
an Environmental Assessment:

- if the site is over 20 ha;
- if there are over 1,000
dwellings within 200 m of
the site;

45

Synopsis of Schedule 2 of the Regulations

Synopsis of the DoE guidelines in DoE Circular 1988/ 15

or gas pipelines, long-distance aqueducts, yacht marinas. (There are also separate regulations for pipelines and transmission lines.)

- if the site is in a very sensitive area, especially if it is linked to other Schedule 2 infrastructure works.

The Environmental Assessment of an industrial estate does not mean that projects in it are exempt from the EA they would incur as individual projects.

(b) Urban Development Projects:

Urban redevelopment does not need Environmental Assessment unless it is Schedule 1 or 2 development, or is very much larger in scale.

New urban development (except housing) may need an EA:
- if the site is over 5 ha;
- if there are over 700 dwellings within 200 m of the site boundaries;
- if the development provides over 10,000 m^2 gross of commercial property;
- if the development is in a conservation or historic area, and with many listed buildings (in which case English Heritage should be consulted);
- if the development is an out-of-town shopping centre with a floor area over 20,000 m^2 gross.

High-rise development (over

**Synopsis of Schedule 2
of the Regulations**

**Synopsis of the DoE guidelines
in DoE Circular 1988/ 15**

50 m) does not need an EA,
but may merit one in conjunction with any other threshold.

(c) Local Roads:
Major roads and motorways
are covered in Schedule 1.

Other roads and major road
improvements may merit an
Environmental Assessment:

- if over 10 km long;
- if over 1 km long and in a
  National Park;
- if over 1 km long and in a
  National Park, SSSI, national nature reserve, or a
  conservation area.
- if over 1 km long and within 100 m of a National
  Park, SSSI, national nature
  reserve, or a conservation
  area.
- if in an urban area and
  more than 1,500 dwellings
  are within 100 m of the
  centre of the road.

(d) Airports
Airports with runways over
2,100 m are Schedule 1. Smaller airports and major works
at Schedule 1 airports may
merit an Environmental Assessment.

(e) Other Infrastructure Projects
Any development with a land-
take over 100 ha may merit
an Environmental Assessment.

ENVIRONMENTAL ASSESSMENT

Synopsis of Schedule 2 of the Regulations	Synopsis of the DoE guidelines in DoE Circular 1988/ 15

6. MISCELLANEOUS DEVELOPMENT:
including noisy or obnoxious projects; holiday complexes, motor-racing circuits, waste water treatment, sludge dumps, test benches for engines, turbines or reactors, scrap yards, cartridge-making, artificial mineral fibre-making, knacker's yards, mine and quarry waste, controlled waste processing below Schedule 1 level.

6. MISCELLANEOUS DEVELOPMENT:
Waste Disposal Installations and landfill sites for ordinary domestic, commercial and industrial waste (not special waste) handling over 75,000 tonnes per year may require an Environmental Assessment. Smaller landfill sites and civic amenity sites for harmless waste should not need an EA.

7. MODIFICATIONS :
Modifications to Sch. 1 projects already completed.

8. SHORT-TERM PROJECTS :
Schedule 1 projects which are mostly for testing and development of new products, and which will only be there for one year.

9. CUMULATIVE DEVELOPMENT:
Not specifically covered in the Regulations.

9. CUMULATIVE DEVELOPMENT:
When a large-scale development is put forward as a series of smaller developments in order to avoid Environmental Assessment, it is for the local planning authority to resolve the matter as part of their strategic planning policy.

Schedule 3: the Environmental Statement

The subjects covered in greater or lesser detail in an Environmental Assessment naturally vary from project to project, but both the Directive and the Regulations set out a basic list of subjects to be studied and these may be supplemented by directions from the local planning authority. This basic list is set out in Schedule 3 to the Town and Country Planning Regulations, but the exact schedule of subjects to be assessed for each project and, as far as is foreseeable, the level of investigation required (the term 'scoping' is really unacceptable) should be agreed with the local planning authority before any assessment is undertaken. This is best done at the same time as the request to the local planning authority for a ruling on the necessity for an Environmental Assessment – that is, right at the beginning of the project. It is not the intention of the legislation to demand a full Environmental Statement on *every* effect listed in Schedule 3, but only on those which are 'significant' for the project under consideration.

The subjects which must be covered by the assessment and included in the Environmental Statement are set out in two groups; the 'specified information', which is mandatory, and the 'explanation or amplification' of specified information, which does not appear to be mandatory although the local planning authority may be entitled to call for it in support of the specified information. Other Environmental Information which may be considered by the local planning authority includes that obtained from the statutory sources given in Chapter 2, and a fuller discussion of the information to be supplied is set out in Chapter 6.

The specified information

This section comprises five subsections which contain the formal statements about the development and its expected significant environmental effects, together with the proposed mitigating measures put forward by the developer.

2(a) Description

A description of the project and its site, and information about its size and design.

Presumably this should cover the site both as existing and as it will be after development. Although not specifically required, information about consequent changes to the immediate neighbourhood of the site should be included.

2(b) Environmental data

The data which will enable the local planning authority to identify and judge the environmental effects of the project.

This data must obviously be capable of standing up to expert technical examination and cross-checking. Environmental Information is best processed in three stages:

1. the raw data such as statistics, questionnaires, survey sheets, traffic counts and photographs;
2. the analysed data giving the summaries of the original findings, their interpretation, when all results are collated and checked;
3. the conclusions as to the extent and significance of the environmental effects.

It is vitally important to keep the data in each stage clearly separated and recorded, and even more essential to ensure that each subject is dealt with equally carefully, although not necessarily in equal depth, since once the opposition can find a fault in the assessment the whole body of Environmental Information becomes suspect. The Environmental Assessment team *must* be able to demonstrate their reasons for every piece of research and how they reached their conclusions from the data available.

2 (c) Significant effects

A description of the probable 'significant' environmental effects on sectors of the environment.

It is important that the developer and the local planning authority agree at the outset of the project which effects are significant, as far as can be foreseen. Much time and effort can be wasted by analysing unimportant effects or, equally, by missing out important effects and having to go back over the ground again. The sectors to be examined are listed in Schedule 3 as:

- Human beings
- Flora
- Fauna
- Soil
- Water   and also the interaction between
- Air   any or all of the first eight sectors
- Climate
- The landscape
- Material assets
- The cultural heritage

2(d) Mitigating measures

The way in which significantly adverse effects can be mitigated as far as may be possible.

This would include both the constructional and operational stages of the project, and covers the design of the project, the operation of the plant, and the methods of construction. The exact nature of the possible mitigation of each effect will vary from project to project, and it is expedient to work out a number of options for these measures, since some may be more acceptable to the local planning authority and the local community than others. The most acceptable option is not necessarily the most expensive, but must be seen to be a genuine attempt to carry out the best remedial or compensating work possible. Mitigating measures fall roughly into four groups:

1 those which are intended to control adverse effects, such as filtering emissions;
2 those which are intended to scale down effects, such as concealing buildings by screen planting;
3 those which are restoration measures such as replacing lost open space or woodland;
4 those which are compensatory (planning gains), such as providing new recreational facilities.

2(e) Non-technical summary

A summary in non-technical language of all the effects listed above.

As mentioned earlier, this is a very important part of the Environmental Statement, since the Press, TV, and public will probably rely on this information for their comments and objections. The summary may be given to a professional writer or publicity manager (under the control of the assessment team) to make sure that all technical points are intelligible to a non-professional audience. As the summary will almost certainly be quoted by journalists, simple and memorable statements can usefully be included where appropriate.

The explanation or amplification of specified information

The explanation or amplification comprises six further subsections which expand the information given in the first part of the Schedule as far as may be necessary to enable the local planning

51

authority to come to a decision on the project. This is supplemented by the additional information obtained from the statutory consultees. The seventh subsection allows the Environmental Assessment team to explain the presence of inadequate environmental information in the Environmental Statement.

3 (a) Physical characteristics

The physical characteristics of the project and its land-take during and after construction.

The basic physical description of the project will already have been given in the 'specified information'; this is probably the place for more detailed descriptions of materials and landscape works similar to those which would be included in the reserved matters of a normal planning application. Most major construction works use land for access and on-site storage in excess of the finished project needs and it is therefore very necessary to distinguish between:

- land taken temporarily, which will be fully restored to its existing or similar uses after construction;
- land taken permanently for the development proper – plant, roads, ancillary structures;
- land taken permanently but reserved for the purpose of mitigating environmental effects – screen planting, replacement recreation areas, public footpaths.

3 (b) Materials and processes

If the project is a manufacturing or processing plant, the raw materials and the processes used should be described.

It is not expected that confidential manufacturing information should be included in this section, but as any competent industrial spy can identify raw material and new plant going into a factory, there is no point in concealing the basic materials and processes. The information is required in order to provide a check on the use of environmentally undesirable materials, including the use of scarce natural resources, and the installation and operation of potentially unsatisfactory plant which may allow the escape of pollutants.

Surprisingly, none of the Regulations calls for any specific statement on the hazards of the project, or how accidents and emergencies are to be dealt with. Many hazards and their controls are dealt with under other legislation, such as radioactive material leakage, oil spills and accidental explosions, but other hazards do not seem to be covered. It would seem sensible for any developer who is aware of possible dangers to indicate his proposals for containing or at least minimizing environmental damage in case of accident. The hazards are not always those due to the process itself; a few years ago a large chemicals factory on the Rhône caught fire and, although normally the materials used in the plant were safely handled, the water used in firefighting naturally mingled with the chemicals and ran off into the river with severe consequences: heavy fish casualties wiped out local fishing, water for domestic use could not be taken from the river and the whole riverside population had to drink tanker and bottled water for several days. Other industries along the river could not draw water, agriculture had to be supplied by tanker, and all bathing was forbidden until the pollution had passed downstream. The final effect on estuarine habitats and the Mediterranean is not recorded. In this case an Environmental Assessment should have revealed the risk, and mitigating measures which might have included bunds round the plant to hold firefighting water could have been incorporated into the design. A parallel case in England, when the water supply for a large part of Cornwall was accidentally contaminated due to poor supervision, might have been averted if the potential hazard of unmonitored access to the waterworks site had been considered as part of an Environmental Assessment.

3(d) Residues and emissions

Any waste product or emission from the project, whether it is considered to be a pollutant or not, including any pollutants which might affect air, water, or soil. Non-tangible emissions such as vibration, light, heat, and radiation are covered by this definition.

This subsection refers to the project when it is in operation; the wastes, pollution, and emissions produced at the construction stage are not specifically included, but there is other legislation dealing with noise, smoke and waste from construction sites. There are particularly strict controls on such wastes as drilling mud from oil drilling operations. In addition, the local planning authority may put some restrictions on site works into the planning permission conditions, or obtain compensation by way of legitimate planning gains.

ENVIRONMENTAL ASSESSMENT

3(d) Alternative development

Any alternative sites, designs, processes or access which were
considered by the developer, but subsequently rejected by him,
together with his reasons for doing so.

The possible alternatives to the siting or access of the project
are usually put forward by objectors, mostly on the 'not in my
backyard' principle, but this is one area where the local planning
authority can be most helpful, as they are in the best position to
know what land is available and how well the project can be
fitted into their development plans. Sites are selected by developers
on the basis of economic viability – the oil company who built
a terminal on Flotta looked at nine possible sites on economic
grounds *only*, with environmental considerations coming in a long
way afterwards – but the many social and political factors which
must form part of a local planning authority's responsibility ought
to be taken into account by the developer before a site and its
access routes are finally chosen.

3(e) Use of natural resources

Although the effects of the project on the environment are described
in the specified information, special attention may have to be
paid to the results of using natural resources, especially where
these are in short supply or are likely to be injured by the
development.

More detailed information on the exact nature of any pollutants,
wastes, or other nuisances (such as noise of plant or traffic) may
have to be provided if their effects are likely to be significant.

3(f) Methodology of assessment

The local planning authority and the public are entitled to know
how the Environmental Assessment team arrived at the conclusions
set out in the Environmental Statement.

The developer may be asked to demonstrate his methods and
calculations, particularly the way in which he obtained an estimate
of the scale of the environmental effects predicted, since the
difference between 'significant' and 'insignificant' effects will
always be arguable. Standard statistical techniques will be acceptable, but state-of-the-art methods, even if superior, may be challenged. For this reason it may be desirable to choose the simplest
and best known methods of assessment available, as long as they

are adequate, rather than more perfect but elaborate methods which are difficult to explain in non-technical terms, and which may not have been adequately tested in the field.

3(g) Difficulties

Any problems or difficulties in obtaining or analysing information should be recorded.

It is not always possible, either from lack of time, lack of knowledge, or lack of basic data, to ensure that all environmental effects are thoroughly studied and evaluated, and any faults or weaknesses in the collection or processing of environmental data should be recorded and justified. No-one can do more than their best to make judgements about possible future effects, and an honest admission that the assessment is less than perfect is more likely to be accepted than an attempted cover-up of inadequate analysis.

Neither the local planning authority nor the consultees want to see complex scientific research, which has little bearing on the practical issues of the Environmental Assessment, carried out for its own sake. Apart from sound research work, many 'scientific' reports show that it is only too easy to drown the real problems beneath floods of technical gobbledegook.

4 Non-technical summary

A summary in non-technical language of all the explanations or amplifications described above.

This complements the previous non-technical summary, and should be carefully cross-referenced with it so that contradictions and anomalies are removed. There does not seem to be any statutory requirement that the two summaries should be separate, and an amalgamation of the two would produce a more readable and reliable document.

Chapter 4

MANAGEMENT

Management structure

No management structure has yet been published which has been specifically designed for the Environmental Assessment of a project, and at present the work is usually carried out by a team of specialists gathered together for the occasion, and most often led by the planning or engineering firm which will take charge of the planning permission submission. For smaller Environmental Assessments, such as those for the more straightforward and single-purpose projects in Schedule 2, most of the work can be done by one firm with the help of specialist consultants for the more esoteric or complicated studies but, for Environmental Assessments like the 'Big One' (the Channel Tunnel), a consortium of consultant firms may be employed. Many larger firms of planners, architects, civil engineers, and landscape architects are now offering a comprehensive service which covers all aspects of an Environmental Assessment but, even so, special advice and assistance is often necessary to provide a complete assessment. Even the comprehensively staffed and experienced Building Design Partnership, which carried out just part of the Environmental Assessment tasks for the Channel Tunnel, consulted 45 organizations in order to cover all possible aspects of their work. There are many disciplines which may be involved in an Environmental Assessment, although it is unlikely that all, or even most, of them will form part of the Assessment team except in the case of major Schedule 1 projects. The disciplines listed and described below provide the environmental skills most likely to be employed in the assessment of a development in a fairly densely settled area.

Town planning will be needed for the study of structure plans, local or unitary plans, and the special emphasis on environmental effects which will occur if the project is in or adjoining an AONB, National Park, or any of the numerous areas which come under some form of planning control. These may include Simplified Planning Zones, Environmentally Sensitive Areas, Conservation Areas, Designated Regions, Designated Areas, National Scenic Areas (Scotland), Green Belts, Heritage Coasts, Areas of

Archaeological Importance, and many locally restricted areas. It may be relevant to note that a third of the land area of the UK is now under some form of restrictive designation as far as development is concerned, and a thorough check is really essential.

Planning law is required for the interpretation of the Environmental Assessment Regulations, and all the other legislation relevant to the project; not least the conflict of Regulations. This discipline will be involved in the conduct of any inquiry, at least until a body of case law and a background of Environmental Assessment experience has been built up in the profession generally.

Property law is necessary for the checks on land titles, easements, and covenants on land. Some easements and covenants can be extinguished, some may be set aside by the courts for one reason or another, and there is an increasing tendency for local planning authorities to 'buy out' covenants from departing landowners in order to facilitate development. An easement allowing the CEGB to run a major existing transmission line across the site of the proposed development, which will then have to be relocated with the provision of easements across other properties, is typical of the constraints on development caused by easements and wayleaves which need to be taken into account. In many linear developments such as major roadworks, the problem of land severance and consequent compensation will come under this discipline.

Civil engineering is needed for the assessment of effects of any major earthworks and structures or land-take which are likely to alter the environment to a significant extent. As the engineer is often the principal consultant for industrial plant, harbour, or highways developments, he is frequently the team leader.

Service engineering takes the responsibility for piped services, their location and possible augmentation or diversion. Local small sewers and other service lines can usually be diverted if the project requires it, but the engineer should be responsible for checking the location of main sewers, major water mains, and electricity transmission lines and underground cables. The development may make extensive demands on the services in the area, and the Environmental Assessment regulations require estimates to be made of the levels of service supply needed. There is one exception to the information collected by the services engineer. Pipelines, their location, depth and contents, are not shown on any map and an application must be made to the Department of the Environment who will say whether any pipelines cross the site, but will provide no further information. It seems that pipelines, together with the position of Regional Government Centres, are a matter of national security and cannot be recorded on local planning authority maps.

In addition, there may be a demand for assessment of sound and lighting effects from plant noise or site lighting, and expert advice is needed to assess the likely disturbance which could be caused by the project.

Architecture is required for the assessment of aesthetic effect of the project on the environment, particularly where the project is close to a Conservation Area or any historic and listed building. The physical design of the project is usually a subject for argument between the developer, the local planning authority, and the public who may be more or less supported by the Royal Fine Art Commission and conservation organizations. Good architecture, sympathetic or challenging to the local environment, can contribute considerably to the mitigation of adverse visual effects. The building surveyor may be called in to advise on the state of existing buildings liable to be modified by the development.

Landscape architecture is necessary for the study of effects on topography, landscape character, tree cover, and visual intrusion. The landscape consultant may also deal with public opinion surveys on visual effects of the project on the landscape. This is especially relevant if the development is likely to impair popular views from look-out points or rights-of-way.

Forestry advice may be needed if an application is being made to the Forestry Commission for a grant in aid of forestry development, or if the scale of landscape planting justifies forestry techniques.

An *agricultural consultant* may advise on the quality of agricultural land and land values in cases where the quality of the land is open to question. He may also deal with the viability of farm units where severance occurs as a result of the development.

Economics deals with the potential job losses or gains, the reduction or increase in local trade and industries, particularly where 'spin-off' service industry or jobs may be involved, and the effect on local rates and land or property values. If there are several options available for mitigating adverse effects, the economist will have to assess the relative economic cost of each. If the client has a financial department or consultant competent to carry out an Environmental Assessment, they may provide this part of the assessment.

Transport/traffic engineering is necessary to calculate the increase or decrease in all types of road, rail and perhaps air traffic generated or affected by the project. The traffic engineer works closely with the civil engineer to assess the location and construction effects of new or enlarged roads, railways or harbour works. Very often the transport engineer is chosen to carry out any noise surveys and forecasts that may be necessary, since compensation or recompense for the cost of ameliorative measures is now payable to residents adjoining newly constructed highways which carry traffic causing excessive noise. Alternative forms of transport – road, rail, or water – may have to be evaluated on both economic and environmental grounds.

Ecology may be needed to examine the effects of the project on the pattern of wildlife in the area, especially if the development is in or near a protected area such as a National Park. Ecology covers the relationship of plant, animal, aquatic, and bird species to their environment, and as there are many protected creatures and plants whose habitat may not legally be disturbed, including bats and natterjack toads, it is advisable to ensure that irrevocable damage to a rare species is not likely to be an effect of the project. The conservation lobby is now so strong that probably only a project of major national importance would be allowed to override the need to protect a valuable ecological system.

Geology may be required if the project is going to disturb any but the surface levels of land, or if the disposal of landfill waste from the project is going to be a problem. In hilly or mountainous districts, the possibility of landslip, erosion or the disturbance of natural rock formations may need assessment. If the project is in a coal-mining area, the British Coal Board geologists will almost certainly be implicated, and if there is a possibility of disturbance to underground waters the geologist will need the additional skill of hydrology.

Hydrology expertise is required to forecast the effect of the project on water catchment areas, underground aquifers, the possible seepage from toxic waste disposal, and the disruption of the natural drainage system of the area. These assessments are particularly relevant where large cuttings, embankments, or excavations form part of the construction process, and also where the natural stream or river system is already at risk from pollution.

Sociology is required for the study of disruption to the local community if this is an especially fragile ethnic or linguistic group, or if the project is going to introduce large numbers of workers with very different cultural backgrounds such as might occur in developments in the Hebrides or the less populated areas of Scotland or Wales. The effect of introducing a higher wage-scale for incoming workers which outclasses the indigenous population, with consequent effects on house prices, cost of living and transport costs should be considered in the Environmental Assessment.

Analytical chemistry is needed for the measurement of existing pollution levels and the estimation of additional soil, air and water pollution which might be generated by the project. Such levels are extremely difficult to measure and forecast, but nevertheless the Regulations are quite specific in their demand for the assessment of future pollution; therefore some estimate, however broad its range, will have to be provided. Because some chemicals which were originally classed as harmless have since been found to be unsafe, it is advisable to consider *all* emissions and discharges from the project and to predict their future effect as far as possible.

Archaeology may be relevant if the project is on a site suspected of containing archaeological evidence of earlier development or cultural remains. It is not easy for the amateur to distinguish the traces of an Iron Age or mediaeval earthwork from modern land-forming, or to identify neolithic artefacts, and an undeserved reputation as a cultural vandal does not help to soothe local public feelings about the project. An expert examination of the site together with a check on historical documentation may ensure that environmental effects do not include any significant archaeological damage.

Public relations are necessary, not so much to assist in the Environmental Assessment itself as to smooth the passage of the planning application through its public participation stages. Except for defence projects, it is practically impossible to keep the news of a major new development proposal secret once the Environmental Assessment has started (someone usually spreads the news as soon as the first thought of the project has emerged from the Chairman's brain), and good public relations will do much to reduce opposition and to direct it into sensible channels. Secrecy may be a necessity to contend with industrial espionage, but it can impair the cooperation of local organizations in the collection of Environmental Information, and invariably makes the public suspicious that they are being conned into an undesirable situation. Local public meetings to introduce the development may be somewhat heated, and even very occasionally aggressive; reasonable discussions are more possible if there is a (more or less) frank presentation of the project and its expected effects. The public relations manager also deals with the media on behalf of the Environmental Assessment team, and thereby prevents unco-ordinated or less-than-tactful opinions from being broadcast before the team has reached its agreed formal conclusions.

Statistics are necessary if the assessment involves the analysis of large amounts of numerical data. Whilst simple totals and averages are adequate for small projects, larger projects may need more sophisticated statistical techniques; small sample analysis, probability calculations, and regression analysis may be required to deal with incomplete or complex data. The French regard statistics as being like a bikini – they arouse interest but conceal reality – but good statistical practice will help to make estimates of environmental effects both reliable and credible. A statistician should be consulted before data is collected rather than afterwards, as the methods of collection may have to be set up in a statistically sound way; if a 10 per cent sample is the minimum required for the accurate forecasting of an environmental effect, then it is a waste of time to collect a 5 per cent sample, or equally, to collect a 20 per cent sample.

Project management as a means of controlling the timetable, budget and employment of consultants is absolutely essential. When the Environmental Assessment is carried out by a consortium of consultant firms, each professional naturally has his own methods of work, and unless the project manager can keep overall control of the programme, the progress of the assessment may not be as rapid or efficient as it should be. Whether the control is exercised by an independent manager appointed for the Environmental Assessment, or by the client's own project manager, or by one of the Assessment team, is not so important. What does matter is the efficiency and tact with which he does his job, and his personality may make all the difference to the quality of the Environmental Assessment and the conviction carried by the Environmental Statement. For very large projects, involving scores of staff, there is some merit in appointing two project managers – one to carry out the purely administrative tasks of budget and programme control, supplies, and general office services, and a professional consultant as a technical manager who is responsible for the quality and quantity of work produced by the team, and who has the authority to override individual consultant's decisions. Such a man might well be responsible for the final preparation of the text of the Environmental Statement, but he will need to command the respect and confidence of the whole assessment team.

The Bar is almost certain to be involved in a large Environmental Assessment if the planning permission application goes to an inquiry, or if the development is dealt with by a Private Bill. Counsel has his own way of looking at evidence – not always

61

that of the consultants or the developer and he should be given plenty of time to study the Assessment before the Environmental Statement is completed. Counsel is not concerned with the facts of the Assessment, except to make sure that he understands the strengths and weaknesses of the consultants' estimates of the environmental effects of the project, but the weighting and pre-sentation of evidence is his proper business, and he may want some alterations made to the text of the Environmental Statement before it goes to the local planning authority. A good counsel is the most useful critic of an Environmental Statement, since he will discover any woolly or unsubstantiated evidence whilst there is still time to correct the document.

Specialists

There are a large number of specialities within the disciplines listed above – economists who specialize in retail trade finance, engineers who specialize in highways, landscape architects who specialize in visual analysis of the landscape, and so on – and it is necessary to employ these specialists if a particular environmental effect requires such expert analysis. University departments are a valuable source of very specialized expertise, and the consultants recruited from universities or major polytechnics are likely to be the best informed experts in their field. They are also usually accepted by the local planning authority and the public as being impartial and unbiased in their evidence, and are well used to summarizing and presenting complex technical information. The ability of experts to speak well in public and to withstand hostile cross-examination, often conducted by skilled counsel, is a great advantage to the client, since even the most sincere and know-ledgeable consultant may be made to look incompetent unless he is experienced in court or inquiry practice.

Monitoring Environmental Statements

There is still a considerable problem to be resolved when the Environmental Assessment has been completed and the Environ-mental Statement is in the hands of the local planning authority for approval or rejection of the planning application. How does a not-very-well-off authority find the money to pay for experts of equal calibre and experience to those of the developer, in order to examine the material contained in a probably enormous volume of data and the resultant findings? Either the local planning authority must take much research on trust, or they must find the skills they need in the marketplace, since however competent the planning staff may be, they have neither the time nor the expertise to check every piece of work presented by the developer. Still less do the objectors to the development have the resources to set up a proper check on the Environmental Assessment, and unless they are lucky enough to have equally competent (and generous) experts in their organizations they are not going to be able to make a fair appreciation of the environmental effects of the project. So far neither the government nor the local planning authorities have put forward a solution to this problem but, unless it can be resolved, the value of an Environmental Assessment will be reduced for lack of competent judgement. This is even more evident in the case of a Parliamentary Committee where the Members, however enthusiastic and intelligent, are not necessarily experts in any of the environmental fields and their judgement as to the validity of the evidence has to be made on the basis of common sense and ordinary lay experience. In the case of the 'Big One', the Department of the Environment employed a full team of consultants to examine and report on the Environmental Statements of the four competing Channel Tunnel consortia, and even then it was not practicable within the time-scale of the project to check every statement submitted; in such a case experience and background knowledge of the area and the regional problems is a valuable aid to checking the validity of evidence.

The checking of information obtained from Structure Plans and other statutory documents, statistics from the Nature Conservancy, the Forestry Commission, and other government and quango sources is comparatively easy and straightforward; the confirmation or rejection of conclusions on predicted environmental effects drawn from the basic data is a much more difficult exercise. The consultant who professes to be a good judge of environmental forecasts needs to have a wide professional experience, an awareness of contemporary and historical environmental happenings, plenty of contacts who can be tapped for unofficial comments, good political antennae, a logical mind, and a great deal of common sense.

Chapter 5

THE ENVIRONMENTAL ASSESSMENT PROGRAMME

It is assumed that by now the local planning authority, or the Secretary of State, or the developer as the case may be, has requested an Environmental Assessment and that the need for it has been definitely agreed. The developer will have been notified of the subjects which the authority consider to require assessment, and some agreement as to the necessary level of investigation (*not* 'scoping'!) will have been reached. It is also assumed that the negotiation and appeal stages have been passed, and that the developer is now committed to carrying out an Environmental Assessment of his project. There are usually considered to be four main stages in the preparation of an Environmental Assessment and Environmental Statement; these can be subdivided into many other sub-stages but, whatever breakdown of the workload is adopted, it is essential to complete each stage before moving on to the next. Not only does this make the project manager's life less hectic, but, because in the nature of environmental analysis every factor interacts with nearly every other factor, it is unwise to leave any section of analysis to be completed later in case it overturns the previous conclusions. The presentation of the data may well be held over until the final stages, but the basic material of the assessment must be established as completely as possible at each stage. The Regulations require copies of various documents to be sent to all and sundry ('sundry' always gets forgotten) at each stage, and the project manager will go crazy if he has to send out amended documents to every one of the original recipients; it is therefore vital to check that each document is quite complete before being sent out.

Stage one: setting up the Environmental Assessment team

Base

The first man on the scene should ideally be the project manager, or the consultant who is to take that role. At the outset, his main responsibilities are arranging accommodation, site access and transport, communications, and equipment if the team has to work away from their own offices, and setting up a central data bank (if this is required) including computer records and hard copy records. He is also responsible for preparing the budget (it is not really possible to allow too large a contingency sum for an Environmental Assessment) and the programme. The Regulations set specific target dates and list the organizations entitled to copies of the Environmental Statement, so in order to avoid omissions and delays one person should be in charge of checking the submission and committee dates, and ensuring that all material is available from the consultants and sent to the right organizations at the right time.

Consultants

The project manager may be in charge of selecting consultants, or at least preparing their briefs and dealing with their contracts and defining their functions in the team. There is no statutory ruling as to who does what in the Assessment team; the division of responsibilities rests with the client, and one or more meetings to agree the provisional scope of the Assessment and the work of each consultant should be held as soon as possible. The range of consultants who might participate in an assessment and their skills has been discussed in Chapter 4. As soon as responsibilities have been agreed (and firmly recorded) the next task is to settle the overall programme and budget with all parties. Each consultant then gets his own team together, working out his own programme and budget, allocating staff to each section of the project and recruiting to fill gaps in his expertise if necessary.

Data control

Each consultant is responsible for ensuring that his material can either be be fed into a central data bank, or made readily available so that other consultants have rapid access to it. This is one of the weakest aspects of most Assessment teams; all consultants *must* be aware, and *stay* aware of the others' work in order to avoid lacunae, anomalies and contradictions which will be the delight of opposing counsel and the media. It is useful to appoint one member of each consultant's team whose additional responsibility is to liaise with his opposite numbers in other teams, and particularly with the project manager. There is much information which can usefully be circulated between team members, and regular information exchange meetings should be held by the liaison staff; these would cover exchange of names and addresses of contacts in various organizations, coordination of interviews to avoid repetition, sharing of hard-to-come-by documents, warnings about uncooperative individuals, and distribution of relevant letters, photographs, and other data.

There are some environmental sectors which will be the responsibility of a single consultant; others will be the joint responsibility of several professionals. Traffic will be dealt with by the traffic engineer; but a lake modified as a result of the development may be assessed as part of an essential water system by the hydrologist, as a recreation amenity by the sociologist, as a visual amenity by the landscape architect, and as a possible pollution factor by the ecologist and the chemical analyst. Such multiple assessments need to be carefully coordinated throughout so that the balance of the environmental effects is agreed between the members of the team. It is very confusing for the local planning authority to be told by one professional that a lake is an amenity for the local community, and by another that it is a potential danger to the ecology, so however true these predictions may be, the conflict of effects should be recognised and reported in the Environmental Statement.

Stage two: survey of existing environment

The existing physical, social, and financial environment must obviously be established before any assessment of future effects can be made. The work can be divided into two categories; fieldwork which includes surveys, trial holes, photographs, and interviews; and recorded data which includes all information obtained from records held by various organizations. Some staff are better at fieldwork and some at paperwork, and it is worth deploying staff to take advantage of their abilities, since a predilection for surveying the local pubs or losing copies of irreplaceable documents is not helpful to a good Assessment. The depth of each survey will vary according to the significance of the effects likely to be generated by the development, so that the effort expended and the amount of data collected need not be the same for every survey, although the level agreed with the local planning authority must be kept in mind, and should any departure from this level occur the reasons must be recorded. It is also advisable to notify the local planning authority and to get their agreement to the revised level.

Fieldwork

Each consultant has his own methods of carrying out his surveys and of recording his information, but as far as possible there should be coordination between consultants as to the format in which data should be presented. If hectares are to be the area unit, all staff should use hectares, not acres; if the limits of many studies are based on local authority boundaries, then all other studies should use the same boundaries, unless there is good reason for divergence. It is essential that all maps should be prepared to the same scale and from the same Ordnance Survey base. In the field itself, there are many opportunities for cooperative working which will save time, money, and aggravation. These include:

- the use of minibuses to transport several survey teams at once;
- exchange of information gathered in passing which benefits another team, or asking another team already on site to check a detail;
- one programme of interviews covering several different subjects – nothing exasperates a hardworking local authority official or busy resident more than a series of earnest young inquirers asking one set of questions after another;
- one or two photographic surveys covering all required views and details;
- trial holes designed to provide all ground information.

These points may seem very obvious, but it is surprising to find how often simple practices like these are not carried out; the hundreds – and even thousands – of pounds which can be spent on duplicated or rechecked surveys could be better used for other assessment work.

Recorded data

The Regulations list the bodies who are statutorily required to provide information for an Environmental Assessment, but they do not lay down any time-scale. It may take some time for an organization – which may be short-staffed – to find, extract and dispatch the required information, and even then it may need to be amplified or explained before it can be included in the Assessment since it is unlikely that the material will be in a suitable form for immediate use. It is therefore advisable to make contact with these bodies as soon as possible, and to specify as clearly as possible what information is needed, and in how much detail. Most of these bodies will have policies on future developments, and possibly research to support the policies, and it may be very helpful to ask for this information in order to assist in the estimation of environmental effects, although these bodies are not legally required to carry out research in order to provide the information needed by the Assessment team. This also applies to those organizations who are not legally obliged to provide information. The date when information was last updated should be checked; some Structure and Local Plans contain obsolete material, and should not be relied on. Even their accuracy should be examined; (one Structure Plan was found to provide water recreation for every single inhabitant of the county, due to a slight error of multiplication by 300 per cent). Useful sources for basic data are:

- rates roll and estate agents — for value of property, ownership of land.
- mailing lists (local authorities sometimes sell lists of ratepayers) — for postal questionnaires.
- aerial surveys which may have been carried out for other purposes — for updating of maps and land uses.
- local history and conservation societies — for unrecorded information on rights-of-way, monuments, etc.
- residents' associations and amenity groups — for gauging public opinion.
- British Rail, bus companies, local transport user groups — for use of public transport.
- schools, parents' groups — for school journeys.

- chamber of commerce, local residents — for shopping patterns.
- local residents, commuters — for private transport problems.
- Land Use Classification maps (very out-of-date) or local planning authority — for land uses.
- Soil Survey of England and Wales — for agricultural land classification
- Geological Survey maps — for geological and hydrological data.
- Mineral Planning authority — for mineral resources.
- water authority — for catchment areas, sewage and water supplies.
- River Quality Survey — for condition of streams and rivers.
- local planning authority and English Heritage — for listed buildings and historic monuments, conservation areas.
- Railway Heritage Trust — for conservation of railway structures.
- local planning authority — for previous planning studies (such as traffic surveys) in the area which may yield useful information.
- local planning authority — for Structure and Local Plans, unitary development plans, local bye-laws, action area plans, and all other designated areas.
- Nature Conservancy Council, Countryside Commission — for Nature Reserves, SSSIs, and other designated areas.
- local planning authority — for AONBs, areas of special landscape value and other protected rural areas.
- other consultants — for previous studies in the area which may be available (at a price) if not confidential.

- RSPB — for bird populations and habitats.
- Forestry Commission — for tree surveys.
- British Coal — for construction in coal-mining areas.
- Department of Transport — for road policies and programme.
- Department of Transport — for aircraft noise preferential routes.
- Gas Boards — for gas mains and supplies.
- Department of the Environment — for checks on pipelines and secret installations (which are not shown on any maps).
- CEGB, Electricity Boards — for electricity supplies and transmission lines
- Her Majesty's Inspectorate of Pollution (Her Majesty's Industrial Pollution Inspectorate in Scotland) — for Presumptive Limits of emissions, air pollution, radiochemical pollution, hazardous waste.
- Marine Pollution Control Unit — for offshore pollution control.
- Central Unit on the Environment, Department of the Environment — for policies on environmental protection.
- Ministry of Agriculture, Food and Fisheries (MAFF) — for levels of radioactivity in water and farmland.
- National Radiological Protection Board (NRPB) — for natural radon levels.
- Nuclear Industry Radiation Waste Executive (NIREX) — for disposal of nuclear waste.
- MAFF — for incineration of waste at sea.
- Institute of Waste Management — for waste disposal techniques.
- Hazardous Waste Inspectorate (in Scotland) — for waste disposal.

These are only some of the most usually exploited sources of information; there are many more sources of specialized information which may be required by the more esoteric members of the Assessment team – it is here that the universities and research institutes can provide valuable data which may not have been published, or is not easily accessible. The Department of the Environment produces various publications describing the work of official and unofficial environmental bodies.

Environmental Information varies a great deal in its quality and therefore in the reliability of the conclusions drawn from it which are used to predict environmental effects. Information may be classified as:

- 'hard' data from reliable sources which can be verified and which is not subject to short-term change, such as geological records and physical surveys of topography and infrastructure;
- 'intermediate' data which is reliable but not capable of absolute proof such as water quality, land values, vegetation condition, and traffic counts, which have variable values;
- 'soft' data which is a matter of opinion or social values, such as opinion surveys, visual enjoyment of landscape, and numbers of people using amenities, where the responses depend on human attitudes and the climate of public feeling.

Participation

The Regulations lay down certain bodies who must be consulted during an Environmental Assessment (listed in Chapter 2), and consultation with them should include discussion on possible ways of mitigating adverse environmental effects, as well as collecting information on the existing state of the environment, since any organization caring for the environment is knowledgeable on the problems and possible solutions in its area of responsibility. Certainly the local planning authority officers should be kept fully informed as to environmental effects that appear likely to influence their policies or programmes, since their cooperation is essential for the satisfactory control of such effects as increased traffic, additional housing, compulsory purchase of land, and re-routeing of footpaths, and they will, quite rightly, object to their forward planning policies being ignored. The successful Channel Tunnel consortium was noticeable for its careful consultation with planning and other authorities and the consideration given to planning policies in the final submission; whether this was the deciding factor in the choice of developer is not known, but there is no doubt that neglect of such consultation can seriously damage the health of the developer's Environmental Statement.

Public participation is another matter. This is a difficult problem; local people and professional environmental watchdog organizations will always be aware of the proposed development and its probable consequences, and it seems better to air the project and to discuss its effects openly rather than to try to conceal the Assessment and its findings. At the time of public participation the project is still being assessed, and the public must be reminded that the final design may change from the original outline submitted to the local planning authority when all the environmental effects and their mitigations have been taken into account. A limited number of public meetings where the project can be shown and questioned, and informal meetings with local groups may be sufficient to bring out any serious objections and to remove misconceptions about the project. Most useful are: examples of previous developments of a similar nature which show how environmental effects have been dealt with, and even offers to take delegates round them to 'see for themselves'; bringing along people from the locality of previous projects to talk about their experience; inviting neutral people of some standing to chair the discussions, and allowing plenty of time for individual discussions with selected members of the assessment team. Models are usually preferred by the public as giving the most comprehensible idea of the project, but as a model cannot avoid showing the whole of the project, it is better to use sketches or simple photo-montages, which do not commit the developer to a complete design. Short hand-outs covering a description of the project and its consequences (not *too* heavily slanted towards the benefits) are useful for those unable or unwilling to attend meetings, and for the local press. Beware of saying too much at this stage; anything drawn, said or written by the team will certainly be recorded and will equally certainly be brought up against the developer at an inquiry or planning committee; most statements about the development should be qualified as being 'provisional' or 'subject to further study'.

Analysis of existing environment

There is a strong tendency to prolong the information-collecting process as far as possible; it is easier than analysing the information, and there is always the hope that someone else will have to do the hard thinking. A firm line has to be drawn at the end of the data collection stage for each environmental sector, which should only be changed for serious reasons such as critical new information becoming available – it must be really critical to the assessment, not just interesting. As soon as all the basic data for all sectors has been collected, the analysis and correlation of information can be carried out. Before the data is completely processed, the Assessment team should agree on the depth of analysis, since if one member goes into great detail, the others will be pressurized to follow; the finer the detail, the more the opposition will look for weaknesses. Oddly enough, a broad statement will be accepted where a minutely detailed piece of work will attract criticism. The quality of the data gathered in the collection stage is important; it is pointless (and dangerous) to use precise analytical techniques on 'soft' or even on 'intermediate' data as the results will have a false appearance of accuracy which can easily be challenged by the opposition. The use of range calculations is very valuable for Environmental Assessment; it is almost impossible to prove a statement that 15.67 per cent of the population use the local common, but practicable and realistic to determine that between 10 per cent and 20 per cent do so, and this more general statement is accurate enough to enable the local planning authority to make a decision on the value of the common to their constituents. Remember that the local planning authority is entitled to know how results were produced, so keep a careful note of the methods used; it may happen that the staff member who did the work leaves without telling his successors exactly what he did, leading to confusion and embarrassment. The methods of analysis may also have to be explained in open court or inquiry, and therefore they should be kept as simple as possible.

Each consultant has his own methods of presenting survey results, but it is preferable to coordinate the use of graphs, maps, and charts by using the same scales, coordinates, symbols and colours for the same items, thus making it easier for the team and the local planning authority to absorb the information.

Stage three: environmental effects

A fuller discussion of possible environmental effects and the mitigating measures which may be taken is to be found in Chapter 6. It is not a comprehensive checklist, but an indication of the types of effect that may be expected in an Environmental

Assessment, and it offers a selection of 'thinking points' which may help the Assessment team to construct their own checklist of effects to be assessed. The exact definition of 'significant' will vary from project to project; an effect which is significant to a small village may well be totally insignificant in the context of a city. Probably the most useful way of recording degrees of significance is to place the effects in ranking order of importance; the obviously major significant effects will be at the top of the scale, and the obviously minor, though still significant, effects will be at the bottom. This method is more effective than simply categorizing the effects as major, minor, or intermediate, as it enables the Assessment team to agree on the significance of effects without arguing which category they belong to. On no account should mathematical weightings be given to the comparative significance of effects; this often leads to a rather arbitrary loading of effects which has a spurious appearance of exact calculation. No two consultants will use the same multipliers for their estimates of significance, and they may find it difficult to substantiate their weightings in an inquiry.

Estimation of effects

Environmental effects can *only* be estimated; not even the most brilliant and far-sighted professional can predict exactly what the final effect of the development will be, and the further the prediction is projected, the less reliable the estimate will be. This obvious fact needs to be reiterated throughout the whole Assessment, both within the team and outwith it. The environment current at the time of the study is the baseline for predicting effects, but it is a basic principle of good Environmental Assessment that the state of the environment in a 'nil' situation should be assessed – that is, the state that the environment would be in if the proposed development did not take place. Only if this is taken into account can a true judgement of the probable effects be achieved. For example, local objectors may maintain that the development will increase the traffic flow to the extent that new roads will be needed, but if the 'nil' situation is studied, it may become evident that other developments already in the pipeline will generate extra traffic even if the project under consideration does not take place, and that new roadworks will become necessary in any case. This 'nil' assessment is very useful when negotiating the level of planning gains with the local planning authority.

A further set of effects may have to be considered if the

development may possibly be abandoned before completion. This event may be due to running out of capital, sudden political or economic changes which make the development unacceptable as it stands, or unforeseen competition from new UK or overseas industries which make the development economically obsolete. If the local planning authority is convinced that such an event is possible, they would probably be entitled to require the developer to provide an assessment of the environmental effects of abandoning the works.

The Department of the Environment warns that if a developer does not give adequate information about possible adverse effects, even after he has been requested to do so, the local planning authority is entitled to assume the 'worst case' for the environmental effect. Even if the effect is really as bad as this it is preferable to submit an adverse effect, together with the suggested mitigating measures, rather than to allow the 'worst case' to be included in the Environmental Statement by default.

The Regulations do not lay down any forecasting time limits for predicting effects, though it seems reasonable to limit the predictions to the life of the project, with the exception of mineral workings where the afterlife of the site is more environmentally important than the operational life of the extraction work. Schedule 1 (Annex I) projects are generally more dominant but not necessarily longer-lived, since changes in policies and technological developments may shorten the life of an installation, while Schedule 2 projects, though less dominant, may have longer lasting effects. This is especially true of projects such as major roadworks or large-scale housing developments which may take years to become fully integrated into the environment, even after they are technically completed. For the majority of developments, it seems reasonable to predict effects for:

- the duration of the construction period, as required by the Regulations;
- five years after completion, which allows time for mitigating landscape works to mature;
- the economic life of the project, amortization period, life of the plant, or lease, since the after-use of the land is an important consideration in suggesting mitigating measures.

The Channel Tunnel is expected to have an engineering life of 100 years,[1] and predictions of effects ought to take that time-scale into account. When long-term forecasts of effects are made it is essential to describe the factors which may upset the predictions,

1 Unless some daft french security guard allows a burning lorry on board.

such as additional housing development, increased unemployment, changes in national or local planning policies, or public attitudes towards the environment, even if these are rather nebulous at the time of the Assessment. Opinions as to what is 'significant' and what is not may change during the progress of the Assessment even though the local planning authority may have agreed a list; it is therefore wise to keep the data and records relevant to *all* effects available at least until the project has received planning permission.

Estimation of indirect and secondary effects

Schedule 3 of the Town and Country Planning Environmental Assessment Regulations calls for consideration to be given to indirect and secondary effects; these are even more difficult to predict than direct effects and such estimates should be given with wider ranges and more conditional clauses. The 'specified information' in an Environmental Statement requires that all 'direct and indirect' effects should be assessed, whilst the 'explanation or amplification' of the specified information calls for 'secondary, cumulative, short, medium, and long term, permanent, temporary, positive and negative' effects to be assessed for the use of natural resources, the emission of pollutants, nuisances, and waste disposal, but not for other factors. This difference reflects current concern for degradation of the natural environment rather than for the impact of the development as a whole. Only the significant indirect effects need be assessed, using the same criteria as those used for the direct effects, and there is a similar need to distinguish between beneficial and maleficial indirect effects.

It is not easy to demonstrate the pattern of relationships between the project and its effects on the environment in words alone, and a plain two-way matrix will help to check that no relationship has been missed out and to explain the pattern to the local planning authority and the public. One matrix may be constructed to show the existing pattern of relationships, and a second may show the estimated direct and indirect effects of the project. Complex computer-generated matrices are not difficult to construct, but they serve little purpose except to cheer up the computer specialist on the Environmental Assessment team: therefore, the simpler the demonstration can be made, the better. The easiest way of presenting information is by way of a key matrix indicating the existence of relationships for each sector (human beings, flora, fauna, and so on) and the relationships between the sectors themselves (water:flora, landscape:cultural heritage). The details of each direct and indirect effect are recorded individually, either on cards or a computerized database. If it is necessary to show short- and long-term effects separately, another matrix will be required. No guidance is given in the Regulations as to the extent to which indirect effects are to be calculated, but the criterion of 'significance' should be used to decide whether or not to include an indirect effect. It is important to make sure that only significant effects are shown, and not every conceivable connection between effects, as this complication would only confuse both the Assessment team and the local planning authority. Demonstration matrices for direct and indirect effects are given in Chapter 7, and these can be adjusted to suit any scale or complexity of development. There is no technical difficulty in constructing practical regional, local, or site scale matrices of effects; the only limits are the team's expertise and the time and money available. A good matrix also acts as a heuristic model, asking questions of the assessor, and tracing the spread of effects throughout the area and beyond.

Proposals for mitigating measures

The Assessment team is obliged to state what measures will be taken to reduce or ameliorate the predicted effects of the development. There is little point in trying to assert that the project will not have maleficial effects when they are obvious, since the objectors will be well aware of them, but the team should also put forward any beneficial effects at the same time in order to show the proper balance of the project. A beneficial effect may not directly cancel a bad one, but a demonstration of the overall balance of effects may make the project more acceptable; it is also reasonable to offset short-term disadvantages by long-term benefits. The most practical way of dealing with this stage of the work is to note the possible mitigations at the same time as the effects are recorded, since the people who provide information about adverse effects are usually those who are best informed about the control or prevention of them, and they may be willing to put their knowledge at the team's disposal. Close cooperation with the local planning authority and other organizations at this stage will help the team to determine what mitigation measures will be acceptable, and to agree provisionally on their implementation, whether this is to be done directly by the developer or indirectly by the relevant authority. It is not much use deciding how a particular bad effect can be offset unless the authority who has the statutory responsibility for supervising the mitigating work agrees; this will only delay the Assessment whilst alternatives are found. For example, it may be possible to mitigate noise from traffic by erecting earth bunds between housing and roads, with shrub planting to improve their appearance, but unless the highway authority are satisfied that the work can be carried out without impinging on sight lines, and that it can be maintained safely, they will not agree to the proposal. The Department of the Environment maintains a Central Unit on the Environment which is a good source of information on the latest policies and technology for mitigating adverse effects. Incidentally, the distinction between mitigating works and planning gains must be maintained at all times.

Stage four: the Environmental Statement

Schedule 3 of the Regulations sets out the information to be contained in the Environmental Statement as discussed in Chapter 3. No format for the Environmental Statement is given in the Regulations, and presumably a suitable format will evolve from experience and with guidance from the Department of the Environment. Today, all the environmental information will almost certainly be held on computer, and it is essential to check that both hardware and software are compatible amongst team members. Data on incompatible systems can be translated by a computer bureau, but this takes time and breaches confidentiality, and whilst the idea of linking all consultants on a network by modem is attractive, this may lay the team open to hacking, which is a risk for confidential projects. Traditionally each consultant prepares his own section of the Statement covering his survey work, analysis and conclusions on the predicted effects within his field. The advantages in this method are:

- The work is completed more rapidly, as each member of the team can get on without waiting for others;
- A particularly good contribution reflects credit on the authors;
- The responsibility for each item can be easily identified;
- Work can continue until the last moment without delaying other team members.

One disadvantage is that sectors may be left out and contradictions allowed in, due to lack of coordination between consultants, and this is a more serious risk in the larger and longer assessments.

The alternative is for all team members to give their results to a central office to be worked up into a single assessment document; the advantages being:

- complete coordination of all results;
- standard presentation and references;
- little or no repetition of site descriptions and background data;
- good cross-referencing of items;
- consistent writing and draughtsmanship.

This method may take more time than individual working, and any extra time should be allowed for in the program.

Whichever method is chosen probably depends more on the personal relationship between client and consultants than on the nature of the Assessment. Like many other multidisciplinary situations, the best solution may lie in a compromise. Each member of the Environmental Assessment team would be responsible for preparing his own professional statement, but within a framework firmly set by the project manager, and with basic information produced centrally.

Individual work would include:

- a report on the existing environment and summary of findings;
- analysis of data;
- an estimation of environmental effects for each consultant;
- suggested mitigating measures;
- preparing Environmental Statement material to a common formula.

Central office work would include:

- controlling the programme and budget;
- preparing base maps and other base documents for all consultants;
- issuing statutory notifications and documents to all parties;
- writing the introduction to the project;
- collating professional descriptions of the site and the development;
- collating professional proposals for mitigating measures;
- final presentation of Environmental Statement documents.

Much of this central office work cannot be done by managerial staff alone, and the technical jobs may be carried out by one or more professional staff seconded from the consultants' offices, or by the consultants' staff themselves working in close cooperation. However the work is done, it is important to make sure that there is a named individual who can be held responsible for every piece of information included in the Statement; not for the sake of passing the buck or for claiming personal credit, but so that any queries can be answered rapidly and reliably. A system of text references in the Statement which can be correlated with the original consultant's work is invaluable. Other consultants, statutory consultees, or the local planning authority who are studying the Environmental Statement, are entitled to ask for elucidation of the documents and it is advisable that all queries should initially be sent to the central office so that they can be recorded formally together with the replies made by the consultant, as these may eventually become incorporated in the final planning submission. An unconsidered reply by an imperfectly informed junior can create a misunderstanding which may take a lot of hard work and fast talking to clear up, and it will invariably concern one of the more sensitive points in the Environmental Assessment.

When the Environmental Statement has been completed and all the loose ends have been tied up, copies have to be sent to the statutory consultees, the local planning authority and, if required, to the Secretary of State; copies must also be available for sale to the public, so that the work of reproduction is likely to be substantial unless the team have given some thought to the efficient reproduction of text, maps, diagrams and photographs. It is not so much the cost of producing elaborate documents, which will in any case be a small part of the total budget, but the delays caused by waiting on printshops and photographic laboratories that makes it difficult to run off copies as and when required. One solution is to ensure that all documents are readable in black-and-white so that copies can be quickly produced on the office copier, and to reserve the more expensive and attractive coloured documents for the statutory recipients. The public may also need to be able to buy cheaper copies, since an economic charge for a large Environmental Statement may be in the region of £20–30.

Remember that the planning committee who make the final decision are neither full-time professionals, nor complete laymen, and if in doubt about the best format for the Environmental

Statement, consult the local planning authority as to the type of presentation which will be most appropriate. The Environmental Statement may be the product of months of hard work for the team, but it is only one of the many important projects that the planning committee have to consider, so some effort should be spent on making it clear, comprehensible, and concise. Very large documents, maps which have to be unfolded on a committee table, or sub-reports which are not clearly related to the master document, are all faults which do not help the submission; they may seem unimportant or even too obvious to mention, but it is surprising to find how many otherwise excellent submissions are physically difficult to handle, and any committee member who has to struggle with recalcitrant paperwork is not in the best mood to consider the project sympathetically. The project manager should make sure that everyone who has to read the Environmental Statement is provided with a good summary which he can carry about with him.

Chapter 6

THE ENVIRONMENTAL STATEMENT: SCHEDULE 3 INFORMATION

'People would die rather than think – and most of them do' (Bertrand Russell)

This is not intended to be a comprehensive checklist of all data needed to complete the Environmental Statement – this would be impossible – but the items discussed may be useful as a mental stimulant for the preparation of an Environmental Statement for an individual project. With many consultants sharing the work of assessment, it is difficult to avoid the repetition of information under different headings; great care is needed to make sure that discrepancies do not appear in the final document when the data on one environmental sector has been revised and the revision has not been cross-referenced to other sectors. The Environmental Statement is written for the benefit of those who will decide the planning application (or any application under other legislation) and the level of technical writing must be set for this audience which is experienced, intelligent, but not expert in each field of the Environmental Assessment, and can therefore make good judgements only if the information is presented in good plain English and, above all, succinctly. Those who make planning decisions have many other duties, and should not be expected to spend valuable time struggling with unintelligible jargon. It should be remembered that both lay and professional people have had little experience of Environmental Assessments as yet, and the Environmental Statement should be presented in a style that makes it easy for readers to grasp the main points of the document, and to understand the developer's proposals.

The Schedule 3 headings in Chapter 3, describing the 'specified information' are taken as the basis for these guidance notes; as each Assessment team gains experience they will build up their own reference material, and as more Environmental Assessments become available for study the range and scope of assessments will improve.

Physical description of the development 2(a)

This comprises a description of the project and its site, and information about its size and design. This section may include 'explanation' material described in headings 3(a), 3(b), and 3(c).

Existing site

- physical survey
- existing land uses
- relevance of any development plans; structure, unitary, local
- any designations affecting the site or its neighbourhood
  - Ancient Monuments
  - Area of Outstanding Natural Beauty (AONB)
  - Areas of Archaeological Importance
  - Conservation Areas
  - Designated Areas
  - Designated Regions
  - Environmentally Sensitive Areas
  - Heritage Coasts
  - Geological Conservation Review sites
  - Green Belts
  - Listed Buildings
  - National Nature Reserves
  - National Parks
  - National Scenic Areas
  - Sites of Special Scientific Interest (SSSI) and others
  (there are also a number of locally designated parks, reserves, areas of great landscape value, etc., listed by the local planning authority which should be taken into account)
- contours and existing man-made and natural features
- existing infrastructure; roads, rail, sewerage, water, gas, electricity

Proposed development

- Schedule 1 or Schedule 2 development
- category in Schedule 1 or 2

- variation from the category
- size, construction, and appearance of buildings
- size, construction, and appearance of installations
- site layout and landscaping
- access to site

3(a) Physical

- land-take for construction period
- land-take for permanent development
- land-take as reserve for future development
- land-take for ancillary development; housing, recreation, and so on
- land-take for amenity; screen planting, screen mounding
- new roadworks for traffic to and from site
- time-scale of development

3(b) Production processes

- if industrial, capacity of plant
- if industrial, processes used
- if industrial, raw materials used
- energy requirements
- natural resource requirements – this is described more fully in section 3(e)(i)

3(c) Estimated emissions and residues from the development

- notifiable pollutants
- emissions to air
- wastes discharged to water
- wastes discharged to soil
- wastes discharged to disposal sites
- special wastes discharged
- noise levels during construction – day and night
- noise levels during operation – day and night
- light levels if night-working
- heat emissions – day and night
- vibration caused by operation of plant
- vibration caused during construction (seismic testing before construction should be included)
- radiation levels – these are described more fully in section 3(e)(ii)

After use of the site

(This applies mainly to mineral extraction.)

- removal of plant and buildings
- proposed land uses after clearance
- time scale of restoration

Environmental data and environmental effects 2(b)

This comprises the data which will enable the local planning authority to identify and judge the environmental effects of the project. It may include the data described in part 3 of Schedule 3, the 'explanation or amplification' of the specified information. Before discussing the likely effects of the development, it is useful to consider what is meant by 'beneficial' and 'significant' effects.

Beneficial effects

Any development is expected by public and planners alike to cause harmful effects on the environment; this is not always, and not even often, completely accurate. Although direct beneficial effects are not so common, there are many indirect effects which confer benefits on the environment, possibly by replacing lost open space by better amenities, or providing social facilities for recreation which were lacking, or even creating new wildlife conservation areas. Most obvious to the public, and most often emphasized by the developer, is the beneficial effect of increased employment, higher wages, and job opportunities for school-leavers. These beneficial effects could be classified as mitigating measures, but properly speaking, a mitigation is intended to reduce an adverse effect, whilst a beneficial effect springs directly from the development.

Significant effects

These are the sectors of the environment for which an assessment of significant effects is required. Not all these effects will be significant for all developments. Each project must be considered individually but, broadly speaking, any effect which pushes the capacity of the environment beyond its normal stable threshold, or distorts the relationships between sectors, should be considered as being significant.

Each professional member of the Environmental Assessment team will be aware of the range of possible significant environmental effects in his field of interest, and of the relevant published research; therefore the catalogue of effects suggested here is only given in very general terms, since every development will vary in its effects and the level of assessment of the existing environment. For example, 'habitat' for wildlife covers all the shelter, water and food requirements of a species (including the chemical and physical properties), and their location in relation to the development. 'Pollution' covers not only the pollutants scheduled as toxic by law, but any substance which harms or even alters the sector of the environment receiving it.

Although section 2(b) refers to the data needed to assess the environmental effects, and section 2(c) refers to the environmental effects themselves, it is more convenient to deal with the description of the existing environment in parallel with the likely effects of the development on that environment, and this is the method used here. Each environmental sector is discussed, first by its assessment, and second by its effects. The sectors given in Schedule 3 are:

- Human beings
- Flora
- Fauna
- Soil
- Water   and also the interaction between
- Air   any or all of the first eight sectors.
- Climate
- The landscape
- Material assets
- The cultural heritage

Schedule 3 of the Regulations is not specific as to the meaning of terms such as 'water', 'climate' or 'the landscape' and the exact definition of these terms will vary with each individual Assessment, depending on the scale and situation of the proposed development. However, there are some basic assumptions as to their meaning which will be valid for most projects, and these are discussed in this section.

Assessment of the human environment

This is the most important sector of the environment which is subjected to environmental effects, not least because people are the only sector that can stand up and protest against development. The human sector of the environment needs to be considered in two ways: as the community in its entirety; and as people engaged in various activities which complicate the ecology of human life. Traditional biological ecology is (comparatively) straightforward since most species only require enough food, shelter, breeding facilities, and security to continue their kind. Man needs all these, and in addition, more chunks of the environment for making consumer goods, playing games, taking holidays, travelling up and down motorways, and other biologically superfluous activities which have to be considered in an Environmental Assessment.

The animal needs of human beings in developed countries are not usually so affected by development as are their economic and social needs, since the basic welfare in health, housing, food, and safety is more or less protected by law and convention, while jobs, recreation and amenities are not safeguarded to the same extent. It is unlikely that any development would immediately threaten the lives or safety of people in the UK, but effects on what is usually referred to as the 'quality of life' are generated by almost any type of development. It is therefore necessary to consider human beings in terms of their activities both at home and at work. These are not environmental in the strictest sense, but are usually included in all Environmental Assessments as a matter of policy. In an Environmental Assessment the same human beings may appear several times:

- as a residential group in a village
- as a skilled trade such as fishing
- as an economic group such as the unemployed.

Each group has different sets of needs and aspirations which may result in contradictory effects.

The main human sectors which are likely to be subject to the most significant environmental effects are suggested as:

- the community – town, village, city centre, neighbourhood;
- population – numbers, age groups, social classes, ethnic minorities, linguistic groups in each community;
- institutions which are mini-communities – hospitals, residential schools, colleges, residential homes, prisons;
- types of employment in each community – unemployed, retired, students, skilled and unskilled workers, trade unions;
- shopping for the local communities and wider catchment

areas – local, sub-regional, regional, markets, size and type of catchment areas;

- education for the communities being assessed and wider catchment areas – primary, secondary, higher, part-time, technical and trade:
- recreation of all types – sport, passive recreation, entertainment, clubs, informal meetings;
- transport, both public and private, and communal – use of local road networks by local and through traffic, access to motorways, parking, rail links, access to stations, commuter patterns, bus services and usage;
- welfare for the communities – emergency services, health services, local authority services, voluntary services;
- existing stresses on the community which may be at the threshold of tolerance for any particular activity or for the community as an entity, such as housing shortage or tourist population pressure;
- land uses as given in the development plans for the area; if the project takes land scheduled for other uses, there may be insufficient reserve for them in the future. In Flotta, the Secretary of State changed the designation of the land from agricultural to industrial in order to make the 340-acre oil terminal site feasible, but this reduction in available agricultural land cannot be made good on an island.

Environmental effects on human beings

Effects on the community as an entity may include severance of sections of the community from each other, disruption of social patterns caused by influx of different populations, the physical, moral or financial domination of the community by development, increased housing demand, and attrition of the community's land reserves. The residents' peaceful existence may also be upset by the intrusion of a large construction labour force, and the noise, emissions, or disturbance of the construction period which may last for several years. One big new neighbourhood development in Scotland took so long to build that the children of the first occupants grew up in the midst of a building site, affected by the unavoidable noise, mud, disregard of the existing environment, and lack of amenities. Other effects may include permanent increases in traffic through housing areas, overloading of services and amenities, and inflation or devaluation of property values.

The population may be sufficiently stable or adaptable to absorb large population changes; if not, some environmental effects might be the alteration of population age or social structure, generating possible conflict or aggression between new and existing groups.

Employment may be subject to effects of changes in employment patterns, which can cause indirect effects such as the type of training skills required, the effect on other employment, and on other industries, and the effect of incoming and construction labour on the labour market.

Shopping patterns may be affected by access to shopping centres, labour shortage, and indirectly by changes in wage rates affecting shopping costs.

Educational needs may be affected indirectly by the development, due to the need for different types of education and training, or by population changes affecting educational needs.

Recreation could be indirectly affected by changes in demand and type of provision for sport, and the consequent need for new recreational buildings and grounds. Recreational shortfall is one of the easiest adverse effects to mitigate, as new sports facilities can be provided by way of planning gains. The informal activities of walking, riding, or birdwatching may be directly affected by land-take or diversion of footpaths, and indirectly by overloading of amenities, or vandalism and disturbance.

Public transport can be subject to effects such as overloading of services, severance of stations and bus services from residential areas, or diversion of roads. Private transport may suffer from road diversions, overloading of existing roads, increased car parking and control, increased damage to roads by development traffic.

Welfare services are likely to be affected indirectly by overloading of existing services, change in demand for types of services due to population and employment changes, and possibly, ill-health caused by pollution. The emergency services are directly affected by the development, since they will have to cope with any human or environmental disasters: although these are not specifically mentioned in Schedule 3, it is advisable to take account of possible hazards and their management in the Environmental Assessment.

Amongst the beneficial effects likely to accrue to a community are improvement of social and welfare facilities, improvements in roads and public transport, and better choices of shopping, education and recreation. Individuals may benefit from higher incomes, increased property values, reduced unemployment or better choice of jobs, and a wider social life. There may be more doctors, dentists, solicitors and other professional services as a result of increases in numbers and changes in age groups in the population.

The local authority may benefit from increased rates, planning gains related to planning permission, and a wider choice of staff and councillors.

Assessment of flora environment

This covers *all* vegetation from giant redwoods to lichens, and all areas from a very small patch of a rare plant to a multi-million hectare rain forest. It is extremely difficult, if not impossible, to separate the environmental effects on flora and fauna from each other or from the effects on the soil, air and water which are the basis of the natural ecosystem, but it is perhaps convenient to catalogue each sector separately even though the effects may be repeated in each sector. *Every* part of the natural world is dependent on every other part, although the connection may appear to be a very indirect relationship, and the assessor should constantly check and cross-check that no significant effect has been omitted. Some sectors which may require assessment are:

- rare species protected by law under the Wildlife and Countryside Act and its Regulations – the schedule of protected species is continually being updated and assessors should check the latest regulation;
- species not necessarily protected themselves, but forming an essential component of an ecosystem (for example, food plants for protected species of birds, insects or mammals);
- commercial forests and woodlands or coppices, both Forestry Commission and private areas;
- plants or trees forming a visual asset to the landscape; these are discussed under Landscape, but their welfare is more usefully considered as part of the Flora sector;
- existing environmental effects which may already have put the flora at risk.

Environmental effects on flora

Direct effects are likely to be obvious ones such as land-take, deliberate or inadvertent destruction of trees or scrub, theft of rare plants, or vandalism by the construction workforce – it was very instructive to watch a motorway repair gang cutting down the carefully planted amenity trees to light fires in the cold weather – and damage to access road vegetation. Indirect effects may include pollution of soil or water, changes in water table or water

90

chemistry (for instance, from alkaline to acid) or changes in micro-climate caused by destruction of shelterbelts or even by large earthworks. Flora can be significantly affected by dust deposition from construction work or operation of plant. Changes in human population numbers or type can lead to vandalism; some species do not regenerate easily after damage or even after overdisturbance.

It is difficult to see what benefits can outweigh potential adverse effects on the local flora from development; the best that can be said is that public concern for the environment is usually heightened by major planning proposals, and that this concern may result in money being found for conservation programmes. Replacement nature reserves and tree-planting by way of screening the development or providing an attractive setting for the project are acceptable beneficial effects.

Assessment of fauna environment

This includes bird, aquatic, and insect life as well as mammalian species. A number of rare or declining species are already protected under other legislation, and the effects on these would almost certainly have to be specifically included in an Environmental Assessment. The effect on local and regional ecosystems is more likely to be of concern to the assessor; it is surprising to note how widely a development can affect the continuance of a particular species or ecosystem. Species which are considered to be pests or vermin have a role in the support of other species and it should not be assumed that any faunal population can be wiped out without adverse consequences. The environmental effects may be considered to be significant if they influence:

- bird, fish, insect and small animal species such as mice and rabbits which can tolerate some disturbance but which are essential components of the ecosystem;
- large animal species such as badgers, foxes, otters, deer who are subject to disturbance of feeding or breeding territories;
- insect species, some of whom play an essential part in commercial pollination or biological control of insect pests;
- aquatic species which include amphibians, fish, waterfowl, snails, and aquatic stages of these;
- rare and protected species of wildlife, which *must* be recorded and assessed;
- migrating species which make use of the land adjoining or

      affected by the development – mostly birds, especially water-fowl;
- existing environmental effects which may already have put the fauna at risk.

Environmental effects on fauna

In most cases, an environmental effect which is adverse to one species is likely to be adverse to others, since prey and predators are closely linked and some species depend on a limited range of foods for their survival. The main difference in effects is probably that between the smaller species of birds, fish, and insects who ignore or tolerate man unless he interferes with their needs, and the larger species of animals and birds of prey who are susceptible to disturbance even without deliberate hunting and trapping. The most obvious direct effect is land-take, either of the shelter or breeding sites, or the feeding territory of a species. Aquatic life is directly affected by changes in water chemistry due to discharges or run-off into streams, pollution, or thermal discharges, destruction of bank habitat, silting-up or flooding due to changes in water management, or changes to the groundwater system. Indirect effects of population changes can comprise overfishing or wildfowl shooting, disturbance of breeding sites, and deliberate dumping of refuse or waste vegetation in streams or ponds, and occasionally vandalism and illegal hunting, pollution of water or food supply, and changes in water supply or location.

The beneficial effects of development are much the same as those for the flora; enhanced awareness of conservation responsibilities, programmes of conservation and protection, and protected areas provided by the developer as a planning gain. Projects which need high security are often extremely beneficial to wildlife – it is a rash deer poacher who seeks his dinner on an American missile base – and the shyer birds may breed successfully in habitats which remain undisturbed for most or all of the year.

Assessment of soil environment

The effects on soil may range from minor disturbance to complete destruction of the soil structure. Small local effects are not usually included in an ordinary Environmental Assessment, but if the effects are likely to be irreversible or to lead to more serious changes in the ecology of a wider area, then even small effects must be assessed. The types of soil which require assessment are usually those which are of agricultural or horticultural value, as land-take of low-grade farming land is not normally considered as a critical factor in seeking planning permission. Other soil assessments may be needed where the soil structure is fragile. The soil types most likely to merit assessment may be:

- agricultural land of Class 1, 2, or 3a, currently in agricultural use;
- light soils which are liable to wind erosion;
- soil types protected by conservation legislation such as natural peat beds, and natural chalk downland soils – it is not the soil as such which is protected, but the whole ecology of the area which may be affected if there is an adverse effect on the soil;
- soil structure adjoining access roads.

Environmental effects on soil

As previously suggested, the effects of a development on soil cannot be readily separated from the plant and animal life it supports. Land-take for the development and its construction account for most topsoil losses. There are two categories of possible effects, physical damage to the soil structure, and chemical damage caused by pollutants.

Heavy machinery during construction may compact soil severely, and it is not always easy to enforce the operations necessary to restore the soil texture after construction. Erosion is a possible effect where shelterbelts or woodlands would be removed by the development or its ancillary works, and the soil is very light. Cut and fill, deep digging for foundations, and piling may disrupt the natural drainage, possibly causing waterlogging or, alternatively, desiccation of fertile soils near the development site; this is a common effect in fields adjoining motorways where culverting has been inadequate. The unauthorized sale of topsoil from the site, and its replacement by less fertile subsoil is a common form of theft which has an adverse affect on the ability of natural and planted vegetation to establish itself after construction, though such sales are improbable on a well controlled site.

Direct pollution from emission fall-out, leaching, run-off, or deliberate dumping of waste, may all affect local soil structure and chemistry, and indirect effects for these sources may have an influence on soils some distance away which are dependent on ground water from the site. As in the case of water, there are chemicals which, though harmless in themselves, may act in combination in soil to produce harmful effects.

The isolation of beneficial effects on soil is a difficult job; the Environmental Assessment team might be better advised to consider effects on soil in conjunction with those on flora and fauna. Such effects might include improvement to natural drainage, removal of existing waste dumps on the site, or importation of topsoil for planting.

Assessment of water environment

Water, being essential for all forms of life, attracts considerable attention in any Environmental Assessment, and except for central urban development, will nearly always be one of the most important sectors to be assessed. The environmental effects on water range from ocean pollution to the quality of tap water, and this section can only touch on some of the most significant effects. The indirect effects of changes to water bodies or systems are often more important than the direct effect; a change in water level may have indirect effects on aquatic life, ground water systems, wildlife habitats, recreation, irrigation, and the interrelationship with streams or ponds. Any given water sector may have to be assessed in several different ways; as a component of the whole hydrological system; as part of a local wildlife habitat; as a source of industrial or domestic water; or as a recreational amenity. The water sectors identified in an assessment could include:

- running (lotic) water – ecology of brooks, ditches, rivers, and streams;
- static (lentic) water – ecology of ponds, lakes;
- wetlands and marshes;
- running or static water used for recreation and sport, fishing, sailing, bathing;
- hydrological systems – aquifers, local ground water;
- natural water supply – wells, boreholes (piped supplies are dealt with under infrastructure);
- running or static water used commercially – fishing, fish-farming, transport, cress-growing;
- water catchment areas;
- reservoirs;
- marine waters and estuarine systems;

- existing discharges which may already have brought the environment to the limit of its capacity to absorb pollution.

Environmental effects on water

The environmental effects of the development on water sectors may be usefully divided into three main categories: the physical disturbance of the water body; the introduction of pollutants or chemicals; and the manipulation of the water flow.

Physical effects Diversion or realignment of watercourses leading to waterside ecology deterioration, and damming and other obstructions which alter the water flow, causing sedimentation or scouring and thus changing the watercourse bed. Alterations to water bodies by excavation or fill which changes the water depths and so the aquatic ecology, and therefore its use for recreation, sport, or commercial purposes. Silt or building waste accidentally or 'accidentally-on-purpose' allowed to enter watercourses or ponds can destroy the balance of the water ecology; even though the impact is temporary, the effects can be permanent. Reservoirs are too well protected to suffer physical damage, but vandalism of banks or stream beds resulting from population changes brought about by the development can destroy plants and animals, besides spoiling the beauty of the area.

Pollutant effects A vast, controversial and ever-growing problem. Pollution in all its forms is one of the most emotive environmental effects, and the assessment of estimated effects in this sector needs great care. Not only may the UK suffer from water pollution generated in this country; there are many sharp eyes and noses on the watch round the North Sea who will very quickly discover any major environmental slip on the part of UK developers who discharge polluted water into waterways flowing to the eastern shore. Water is terribly vulnerable to apparently small alterations in its constitution, and much of the damage is irreversible. As with all other continuous operations, it is the cumulative and surge effects of discharging effluents into water systems which can produce the most significant effects. The obvious toxic pollutants, oil, and heavy metals such as lead, cadmium and mercury, are dealt with by other legislation, but any substance not naturally present in the water is a pollutant in the strictest sense. These include run-off fertilizers, pesticides, alkaline, acid or other effluents which change water chemistry, abnormally hot or cold water, reducing agents which diminish the oxygen content, and lethal compounds of otherwise harmless separate discharges. Marine and estuarine waters are vulnerable to changes in salinity caused by altered river discharge. The way in which effluents are discharged into a water system is also important, as concentrations can unbalance the water ecology. Sewage effluent, even when properly treated and discharged can overconcentrate nutrients to the point of eutrophic destruction of the aquatic ecology.

Effects on water flow Effects on water flow may be produced by taking water for industrial processing or cooling, diverting watercourses, or changing the ground water system by excavation or other earthworks. Reduction in water supply is the most obvious effect and probably the most serious to commerce, wildlife and recreation, but an irregular flow causing alternate drying-up and flooding is almost equally disastrous. Trees are especially sensitive to changes in water supply, and the results of water shortage may not be evident for some time, when it is too late to take remedial action. Even fairly small changes in water level which occur more rapidly than normal seasonal changes can upset the ecology of wetlands and marshes, and the viability of commercial water uses such as fish-farming, irrigation, or boating. More far-reaching, and much harder to establish, are the changes to the regional hydrography caused by disrupting waterbearing strata or the natural drainage system; these may affect far distant water quality and quantity. Authorities who used to draw public water supplies from deep arterial wells have had to stop doing so because of pollution from aquifers many miles away.

Beneficial effects on water are not so much constructive as remedial. They comprise the reduction of pollution or effluent damage because of improved processing of raw materials by a new development which handles discharges more safely than the previous plant, or enforced compliance with environmental regulations as a planning condition on extensions to existing installations. Planning gains may be used to improve water bodies under the control of the developer, and land drainage of the site may help to improve the natural drainage system if it has deteriorated.

Assessment of air environment

This refers to atmospheric pollution as a rule, since any other type of air damage is, fortunately, beyond the capacity of current technology. The UK does not often suffer major environmental damage from air pollution by other countries, since the British Isles are usually to windward of continental emissions and fall-out. However, we can suffer from emissions and radiation that have travelled round the globe, and from our own local pollution within the island boundaries. With the further integration of the UK into Europe, continental powers will be better able to enforce environmental controls and standards on those of our industries which cause significant effects, and it is likely that the assessment of these effects will need more study than has previously been the case. Air and water are the two possible carriers of pollution which may affect other countries more severely than our own, and a good look at the downwind environment abroad may be an essential part of an industrial assessment. The same sectors are vulnerable both in the UK and on the continent, but their location should be clearly stated in the Environmental Statement. They may include:

- communities near enough to the development to receive significant amounts of pollutants;
- scientific establishments who rely on clean air such as observatories;
- tall chimneys serving industrial processes;
- existing sources of air pollution which may have already brought the environment to the threshold of significant effects on the ecology of adjacent or remote ecosystems.

Environmental effects on air

Estimating the quantity and toxicity of emissions, let alone the likely effect on human, animal and plant life, is an extremely difficult task. The best that can be done is to study earlier developments of the same type which use the same processes (if there are any) and to project forecasts based on that information. The direct effects on air itself as atmosphere and stratosphere are probably limited to long-term repercussions on the ozone layer and similar effects. Thus, the effects which are most likely to concern the assessor are those of dust deposition on flora and water, toxic emissions which cause ill-health or genetic interference, and irritant particles which cause discomfort and affect delicate mechanisms. There are also emissions which can change the chemical balance of rainwater – acid rain is the best known example of this effect, and somewhat similar is the reduction of sulphur compounds in London air which is said to be the cause of increased pest attack on plane trees. One type of effect which creates powerful opposition from objectors is the smell produced by certain processes or types of farming even though the odour may be quite harmless; the argument usually put forward is that property values may be depreciated by evil odours permeating houses and gardens.

Beneficial effects on air are much the same in principle as those on water; they amount to a clean-up of previously harmful emissions by the introduction of new plant or improved processes, rather than any positively beneficial effects such as can be expected from physical alterations to the environment.

Assessment of climate environment

Macro-climate This definition covers meteorological conditions recorded for areas over 100 sq. miles, and this sector is not really relevant to UK Environmental Assessment, since our national climate is determined by factors beyond our control. Overseas, some fairly well established local or global climatic sectors are:

- changes in rainfall pattern, possibly caused by very large-scale felling of tree cover;
- changes in the world atmosphere – this is the basis of the potentially disastrous 'ozone hole' environmental effect;
- desertification or major changes in vegetation;
- unknown climatic changes caused by large-scale changes to water bodies, such as the Russian proposals to divert a continental river and to drain an inland sea;
- major radiation leakages that may possibly affect the climate.

Meso-climate This refers to the local climate recorded for areas under 100 sq. miles and can be determined for smaller areas where the topography and vegetation are clearly seen to be climatic influences. This is especially true for large conurbations which may be slightly changed by multiple developments.

Macro-climate is not normally considered as part of an Environmental Assessment. Even the meso-climate is not likely to suffer permanent significant effects from development unless it is enormous both in scale and in its polluting powers. Developments which may merit some examination might be:

- those which generate high thermal outputs such as cooling installations or large industrial complexes;

- high developments which can produce 'wind-tunnel' effects or heat build-up on the immediate climate;
- developments in areas of unstable climate which may result in dense traffic congestion – such as downtown Los Angeles where 'smog' could have an indirect adverse effect on health in the locality.

Environmental effects on climate

Difficult to assess, especially for one individual project. Major industrial development may cause some changes in local climate, but the effects which need consideration are the more local ones of increased wind pressure on pedestrians, vehicles, and buildings due to high and dense development, and increased thermal output from buildings which may affect other developments close to the site. It is possible for 'smog' to cause illness and deaths in people already at risk from heart and lung conditions, and the reduction of sunshine levels due to continual emissions could affect plants in the immediate vicinity, but it is nearly impossible to attribute these effects to a single development, however large.

Beneficial effects are again likely to be those of reducing thermal output and wind-tunnel effects of previous development; they are remedial benefits rather than constructive ones.

Assessment of landscape environment

Physical landscape This definition covers the whole range of physical components of the landscape, which may be taken to refer to the townscape if the development is in an urban area. The range includes topography, skylines, woodland, specimen trees, smaller vegetation, water bodies, man-made features, hedgerows, land uses, and other physical components as they affect the landscape prospect. Both the landscape of the site itself – if it is large enough to incorporate landscaped areas – and of the surrounding area may be affected by the development. The topography of the area is an important consideration, since the design of the development may be able to take advantage of ground forms to reduce the visibility of buildings or plant.

Aesthetic landscape This covers the human appreciation of the landscape as distinct from the individual landscape components. It is extremely difficult to assess satisfactorily, as the visual quality of the scenery is so much a matter of individual taste and opinion.

The more magnificent landscapes are not likely to be subjected to development, since most of them lie in protected areas, and are the main reason for those areas being protected; but it is the marginal land where the landscape is attractive but not outstandingly beautiful that the assessment becomes most difficult. This is an area of assessment where the scope of studies may extend well beyond the site and its immediate environs, as views and scenic routes can be impaired by comparatively distant development.

The principal landscape sectors which may require assessment could be:

- land uses which are usually considered under the heading of landscape;
- protected areas such as AONBs or National Parks which will already have been assessed for their planning implications, but should also be assessed for their landscape quality;
- geology and hydrology of the site and of the area as far as they affect the landscape;
- topography of the site itself and of the surrounding area;
- tree cover and vegetation (hedgerows, scrub, grasslands) which is an essential component of the local ecology;
- watercourses and water bodies as far as they contribute to the landscape;
- traditional cultivation patterns; stonewalled fields, double hedgerows, and coppices which contribute to landscape character;
- landscapes with historic or other associations.

Environmental effects on landscape

Any effects on landscape are bound to be almost inseparable from effects on soil, water, flora, and fauna, whilst the human factor is important to the evaluation of effects on the aesthetic landscape. Even the most megalithic development is unable to affect the geology of an area, but the topography may well be changed to a certain extent by very large earthworks such as those which are created by the construction of a dam, motorway, or airport runway. The most conspicuous, and therefore the most controversial, effects are the removal of individual trees, woodlands or other vegetation which change the character of the landscape, but the equally significant, though unnoticed, effect of allowing vegetation simply to die of neglect, starvation or old age is not recognized as an environmental effect. Planning conditions may require that trees should be preserved, but the continuation of such preservation is not enforceable; well-landscaped factory grounds have been seen to relapse into a rubble strewn second-hand car dump in two weeks. There is no reason why the mitigating measures approved by the local planning authority should not include lifetime maintenance and replacement of trees and other vegetation.

One of the local community's most sincere objections is to the despoliation of beautiful aspects or prospects of landscape in their area. This need not mean that the development is stuck right in the centre of a classic view; even the suggestion of industrial or commercial intrusion into a rural or historic landscape can create a significant effect – imagine the impact of even the simplest hamburger bar in the Pass of Glencoe – and this is one area of Environmental Assessment where the possibility of re-siting or redesigning the project should be carefully considered. It is not realistic to assess the landscape from every viewpoint, and it seems reasonable to consider the effect of the development as seen from public roads and footpaths only, or from well established viewing points.

Assessment of material assets

This term could cover almost any physical or indeed, any non-physical sector that could be said to have a material value. The proof that a development is likely to have an effect on anyone's assets probably lies with the claimant rather than with the local planning authority or the developer, and it is very improbable that both sides in a planning dispute will agree on the value of any property or asset. No provision is made in the regulations for arbitration for any dispute as to valuation, and the most likely source for an acceptable valuation is the District Valuer.

The value of intangible assets is less likely to be agreed; the impact of development on 'unspoilt countryside' advertised in tourist literature may cause material losses to local hoteliers and shops. Significant effects may require an assessment if the development:

a) debases property values by noise, traffic, or nuisances;
b) increases transport costs due to severance by road or railways;
c) causes loss of trading income;
d) causes deprivation of communities – loss of schools, social centres;

e) severs land ownership parcels – farms, institutions, public open space, commercial woodland;
f) compulsorily acquires land in public or private ownership at values lower than the market price;
g) sterilizes mineral resources of value; if these consist of oil, gas, coal, or any other Government monopoly mineral, then objections to the development will be considerable.

Effects on material assets

Obviously, the depreciation of property values is in the mind of the majority of objectors. There are several different types of depreciation which need to be examined; the loss of house values caused by the near proximity of 'bad-neighbour', incompatible, or simply unattractive development; the similar loss of value affecting recreational or commercial interests, and the straight-forward loss of income from competition by rival companies. These effects may be imagined in combination if a mega-store were to be developed in a top-of-the-market tourist district which is served by small retailers. It is very difficult to suggest mitigating measures for material asset effects since any compensation to residents or to an existing rival firm will not be entertained by the developer. A typical example is the proposal to construct additional tourist accommodation in the Scilly Isles; the islanders have been deprived of their traditional occupations of early flower growing and fishing by international forces and have turned to tourism as the only economic alternative, but the very accumulation of hotels and holiday camps would destroy the touristic value of the area. The sterilization of mineral resources by building on top of them is another material asset difficult to measure, since the extent and quality of the minerals is largely unknown, and few developers would yield to claims that they are sitting on goldmines and should therefore compensate the nation for the loss of revenue.

There is often a conflict between the developer and his backer and the local population on the question of beneficial effects. Even though the development may increase jobs and wages, a large part of the profits from the development will go to distant shareholders or conglomerates; and proposals for spin-off or chain developments which would benefit the local community are worth considering as part of the main project.

Assessment of the cultural heritage

Presumably these sectors will only legally merit assessment if they are already scheduled by a conservation body, or local planning authority. There are many areas however, particularly those where archaeological evidence is known to be present but is not visible on the surface, which are not very well recorded or protected, and it is up to the local societies to make sure that all such areas are at least made known to the local planning authority before the development is approved. It is not so long ago that an area known to contain some Saxon burials was written off, quite reasonably, as unlikely to be of great archaeological interest, yet the local society ran a rescue dig which produced an extremely rare and perfect specimen of a very fine early glass bowl.

On a larger scale, fashions in conservation change with the generations; the Georgians rebuilt 'horrible damp dark old castles', the Victorians re-faced the 'plain-looking' Georgian buildings, the neo-Georgians removed the 'hideous Victorian edifices', and we are busy demolishing the housing and town centres of the 1930s. Both local planning authority and developer need to consider the possible conservation of any building which is good of its kind, regardless of fashion, since a well designed and constructed building is always an asset to any culture, and current technology is capable of preserving and adapting almost any structure. Any assessment of the cultural heritage should take all these aspects into account. Effects on the cultural heritage may be significant if the development impinges on:

- a conservation area;
- an historic town or town centre;
- an ancient monument – obviously no development can take place *within* a monument, but development may be prohibited within eyeshot of the site to protect the ambience;
- special historic or cultural landscapes as described above;
- areas of archaeological importance.

Environmental effects on the cultural heritage

Effects range from total destruction of buildings or archaeological sites to simply spoiling the setting of a historic building. The present instance of the battle for London's skyline is a typical example where one development may slip through as not producing a significant effect, but where a number of buildings can completely ruin a fine prospect. Public awareness is a big factor in deciding the significance or otherwise of a visual effect; so is the proximity of the next election. It is difficult for the Assessment team to appreciate the values of cultures different from their own, even if at the other end of the same country, and therefore bodies such as English Heritage, the Countryside Commission, the Civic Trust and the Royal Fine Art Commission should be consulted, as they are considered to be the accepted judges of the scale of effect that a development may have on a historic townscape or landscape. Many cultural assets are held by the National Trust and their estimate of effects should also be taken into consideration.

Effects on the infrastructure

Although the Regulations do not specifically call for assessment of effects on the infrastructure, since this is usually dealt with under the normal planning permission procedures and is not strictly part of the environment, any major development is bound to affect the pattern of roads, railways, and piped services. The project may not require new access roads, sewerage, or rail links, but the increased use and possible overloading of the existing infrastructure will probably have knock-on effects on the environment. For instance, a road which could previously be crossed without danger may become a serious hazard due to heavier traffic generated by the project, requiring pedestrian and traffic control, school buses, and extra policing, all burdens which fall on the local authority (and the ratepayers) rather than on the developer. Additional housing for incoming workers may take up all the spare sewerage capacity, leaving the next housing area to support the cost of new main drainage. Whilst the developer is not expected to emphasize these points, it is advisable to include them in the Environmental Assessment, partly because the Regulations do call for indirect effects to be assessed, and the local planning authority can therefore demand this information, and partly because it is easier to negotiate the mitigating measures required to offset the infrastructure loading *before* the project goes to the planning committee.

Hazards

There is no requirement in the regulations for the developer to discuss the type and probability of accidents in his plant, although many of the possible hazards and their management are covered by other legislation which deals specifically with radioactive leaks, oil spillages, explosions, toxic chemicals, and prevention of accidents at work. However, it is good practice for the potential dangers of a major development to be considered in the Environmental Statement, and the hazards evaluated along with the various options for their control. Every development will have different hazards according to its nature and siting; what may seem an unimportant minor accident in an industrial process may mean death from pollution to part of the local ecosystem. Many hazards can be simulated by computer, and though this may not include all the side-effects of an accident, it does give some basis for designing preventive measures. Some of the most commonly found hazards are:

- Persistent seepage of radioactive material, toxic chemicals, effluent from waste stores, and other faults in the installation. These may be caused by negligence, inadequate maintenance, or sabotage, and the Environmental Statement should set out the methods of supervision and monitoring they propose to use. This is a difficult provision to enforce, since the local authority has neither the time, manpower, or expertise to check the firm's monitoring procedures and even the statutory authorities responsible for control are sadly understaffed. It has been reported that each inspector for health and safety on building sites in London is expected to supervise some 17,000 sites – an obviously impossible job.
- Disasters: dam bursts, explosions, fractured main pipes, fires, earthquakes (yes, there are earthquakes in the UK), collisions, processes out of control, structural failures, or other unmanageable happenings. These are the hardest hazards to foresee and to control, but the developer can at least make some provision for dealing with the known emergencies in his trade, and also for warning the local community if they need to run for it. Victorian factories and mines often had a system of warnings, given by the factory hooter, for alerting people to fire, flood, or explosion. Good security will help to prevent disasters from being engineered by terrorists.
- Offsite hazards: these include dangerous traffic to the site, illegal dumping of hazardous or special waste by lazy or dishonest contractors, road accidents involving explosives or dangerous chemicals in transit to or from the plant, and washing out tanks illegally. It is true that the developer is not strictly liable for other people's misdeeds, but nevertheless

he will be held to blame by the local community and should therefore make an attempt to evaluate these hazards and minimize them as far as he is able.

Adverse environmental effects during the construction period

In addition to the adverse effects already discussed, there are a number of other effects which are only present during the construction and commissioning stages of the project. These range from massive land-take for casting yards and temporary access roads to the socially disruptive effect of a large active workforce in the neighbourhood. The land-take is unavoidable, but precautions can be taken against leaching of chemicals from dumps and oil spillages from plant and fuel stores. Oil exploration firms have to construct bunds round the entire site, and this method does ensure that no pollution leakage occurs. A good deal can be done to reduce objections by good information provided *before* construction starts, setting out the annoyances that will inevitably be caused, explaining how these are going to be reduced as far as possible, and giving the timetable of the work so that the nuisances can be seen to have an end. Apologies for nuisances, and offers to carry out small remedial works will help to reduce aggravation; their cost will be a very small percentage of the project budget. Some typical construction works likely to cause adverse environmental effects may include the following.

- Damage to adjoining roads by site traffic can be reduced by quick repairs and checking truck speeds. Mud and debris from site traffic on roads must be cleaned up by law anyway, but extra care should be taken in residential areas.
- Noise from construction sites is controlled, and controls may be included in planning conditions; extra provision may be required if work is carried on at night and weekends, especially near institutions where the occupants cannot leave the building. Pile-driving is one of the worst noises, since it travels through structures and cannot be insulated; it is desirable to limit pile-driving periods to daytime and weekdays if at all possible.
- Firm policies for the regular organized removal of building and domestic waste from construction sites should be established as, although there are regulations governing site fires and rats, these are often not enforced, to the fury of the

local residents who object to their homes being blackened or gnawed.

- Danger to pedestrians from site traffic, particularly on rural roads without footways, can be met by prohibiting site traffic at rush hours and school times and controlling lorry speeds.
- Damage to woodlands, wildlife habitats, and adjoining protected areas may be reduced by disciplining the would-be poachers in the workforce and good security fencing which may become part of the final fencing contract. Individual trees need JCB-proof protection and may need watering during the construction period.
- When development continues over many years it may be necessary to provide temporary shops, meeting rooms, day nurseries and other social needs of local residents, which are severed by the construction work. These may be mobile buildings or even stances for mobile shops and libraries. For example, a large new road complex may involve minor public roads which will remain open after completion, but which are closed during construction, forcing the local population to go a long way round to their former amenities.

Direct and indirect effects

Both the Directive and the Regulations call for an assessment of the 'indirect' effects of the development on the environment – namely those effects which result from the action of the direct effects on environmental sectors. Typical examples of indirect effects would be the loss of fishing licence revenue caused by the direct effect of river pollution on the fish stocks; the change in shopping patterns caused by the direct effect of a change in the age-range of a new workforce; the disruption of social and shopping links caused by the direct effect of diverting public transport; the vandalizing of a woodland caused by the direct effect of re-aligning a right-of-way. Indirect effects are much harder to identify and to assess; there is a strong tendency to leave them until last in the Assessment schedule and to hope that full evidence on them will not be demanded. It is certainly possible to make some estimate of their importance and severity, and even to assess tertiary effects, but this last exercise is only really necessary when a tertiary environmental effect is critical to the success or failure of the planning application – a situation which is very unlikely to arise in practice.

There is no agreement amongst Environmental Assessment

authorities as to the range over which environmental effects should be calculated. At present, the practice is to calculate effects for the site and its immediate environs only since these are the only areas which are under the control of the developer and can be provided with mitigating measures. The best example of this is the construction of a generating station, where the CEGB confine themselves to the station itself and its effects on the surrounding landscape and the immediate infrastructure, but the open-cast coal mine which supplies fuel, the railways bringing coal and removing fuel ash, the fuel ash processing plant, the transmission lines from the station, and the distant estuary which receives the cooling water, all fall outside its control and therefore its proposals for mitigating measures. The controversial location of the rail link from the Channel Tunnel is likely to make a very large number of people aware of this discrimination between the project itself and its more remote consequences, and it may be that the fight between British Rail and the local communities will bring forward official policies for dealing with this problem, or different ways of solving it.

Mitigating measures 2(d)

These comprise the measures required to mitigate adverse environmental effects of the development.

A clear distinction should be made at the outset of the Assessment between mitigating measures and planning gain. Before any mitigating measures are put forward, the developer and the local planning authority must agree as to which effects are to be regarded as adverse, or sufficiently adverse to warrant the expense of remedial work, otherwise the whole exercise becomes a bargaining game which is likely to be unprofitable to both parties.

It is probably most useful to present the mitigating measures in parallel with the assessments and significant effects already discussed. A four-column layout is often used to show the existing environment of each sector, its assessment, its environmental effects, and the relevant mitigating measures side by side. With a little bit of luck, work which is required to mitigate some effects at the construction stage may be used as permanent amelioration. It is impossible to be specific on every possible type of mitigation, but some of the more commonly used ways of mitigating effects are discussed here. They are divided into four groups:

1 control of adverse emission effects, such as filtering and detoxification;

2 reduction of adverse visual effects, such as concealing buildings by screen planting;
3 restoration measures such as replacing lost open space or woodland;
4 compensation measures, such as providing recreation or social amenities (planning gains).

Group 1: direct prevention

This includes all preventive measures taken at source to stop emissions, wastes, smells, and other discharges from leaving the installation. It is possible to produce genetically engineered bacteria which will deal with almost any toxic chemical, but this apparently ideal solution has science-fiction possibilities for disaster should any of these unnatural organisms desire a change from their restricted diet of poisons, and start to invade the outside world. More conventional solutions include:

- Optimum technology prevention of non-toxic emissions by filtration and cleansing systems.
- Toxic emissions are already dealt with under separate legislation, but additional precautions may be desirable.
- Radiation and radioactive fall-out are covered by other legislation, but a clear statement of the precautions taken may help to satisfy the local planning authority.
- Controlled liquid waste disposal by detoxification or removal from site.
- Controlled solid waste disposal by regular removal from site (special waste is already covered by other legislation). Leaching of effluents from storage lagoons or dumps is a very common hazard of chemical or agricultural processes.
- Sound insulation of buildings and plant where practicable, or attenuation of noise by screening structures.
- Attenuation of smell by filtration.

Group 2: reduction of adverse effects

This includes design and construction measures to reduce the visual effect of the development and may involve:

- creating superbly designed structures which are an asset to the environment;

- siting as much of the project below ground as is feasible – possibly plant rooms, storage, or car parking;
- designing the project to take advantage of topography, either to screen the buildings or using building forms compatible with the landform.
- screen planting of tree cover and vegetation – this can be either immediately adjacent to the buildings or installations, or in strategic positions in the surrounding landscape in order to direct the eye away from the development;
- designing low-key structures – simple profiles, vernacular materials, muted colour schemes, or domestic-scale groups of buildings – rather than mega-structures where possible;
- reducing ancillary works to a minimum – undergrounding cables, enclosing and screening store dumps and waste dumps;
- shielding working lights used at night – modern lamp designs allow good directional control;
- careful alignment of roads and yards to prevent traffic noise from disturbing the local community. Acoustic barriers (earth or concrete) can help.

Group 3: restoration measures

These are intended to replace any environmental sectors lost to the development – perhaps not exactly, but sufficiently near to the original to fulfil the same function. A complete list of all possible losses would not be feasible, but some types of environmental loss are fairly common to all developments. A lost or damaged sector may not be replaceable on site if there is not enough room: there is thus a good case for *increasing* the land-take of the project in order to leave room for restoration work.

- Woodlands may be replaced with new planting of similar mix on site or on extra land-take. A number of small woods are *not* the same as one large wood as far as wildlife is concerned.
- Specimen trees may be replaced in positions where they will enhance the landscape; a new classic entrance avenue of limes, oaks, or chestnuts is an example. Protection from vandals is essential.
- Water bodies may be replaced with new ponds or lakes, but they must be capable of developing into natural water bodies and care must be taken that they are compatible with the local water systems and ecology.

- Streams or brooks which have been diverted should be of similar flow patterns to the original; culverting should be restricted to short lengths interspersed with open sections.
- Public open spaces such as playing fields, recreation grounds, and playgrounds are fairly easy to replace, and the new situation may be an improvement on the original. Allotments and well planted parks are a more difficult problem, as it takes some years to re-establish crop-growing land and mature shrub and tree planting.
- Severance of community links may be overcome by providing parallel facilities for the separated part of the community, such as bus links, meeting halls, playgrounds, corner shops, doctors' branch surgeries.

Group 4: compensation for adverse effects

In many cases it is impossible or impracticable to replace a lost environmental sector exactly, and the local planning authority may accept that compensation in the form of alternative benefits can be made. Each authority will have different environmental needs which can be supplied by a willing developer, but it is advisable to match the compensation as accurately as possible against the environmental loss; lost playing-fields should be made good by some form of recreational provision and not by a car park for the town centre. Planning permission often includes conditions requiring the provision of planning gains by the developer to offset some deterioration of the area caused by the development, but it is essential to distinguish very clearly between those benefits offered by way of compensation for adverse environmental effects, and those which are a formal part of planning consent. The local planning authority may decide to formulate the compensation proposals as a planning condition in order to ensure that they are carried out, so the developer should beware of putting forward proposals that he does not really intend to implement.

Some possible compensations for adverse environmental effects are suggested below.

- Increased traffic and noise: by paying for improved road junctions, pedestrian crossings, and providing acoustic barriers in excess of the legal minimum.
- Loss of public open space: by buying land elsewhere and providing similar recreational facilities there.

111

- Loss of amenity trees and other vegetation: by planting trees in the community in selected areas.
- Visual intrusion by the development: by subsidizing screen planting by individuals and organizations around the site, re-routing footpaths and viewpoints whilst improving their condition.
- Loss of wildlife habitat: by buying land elsewhere to re-create a nature reserve or bird sanctuary area.
- Enhanced or decreased property values: by offering compensation for loss of value. This is not really practicable even if it were desirable, but building low-cost starter homes, advance factory units, or retail units to compensate for rising property prices is socially acceptable and may even be profitable.
- If no other form of mitigation for adverse environmental effects is possible, it seems reasonable that the developer should offer to support or subsidize existing or proposed environmental projects in the community, especially if these are seen to be educational. Sponsoring environmental clean-ups, school visits to wildlife areas, scholarships or fellowships in environmental sciences, or research into ecology or other natural sciences, would appear to be a useful gift to society in return for the development.

Summary 2(e)

A summary in non-technical language of the information specified above.

It is to be hoped that the definition of 'non-technical language' includes a ban on all jargon, which confuses the lay public even when, on rare occasions, it is not intended to do so. The risk of using confusing and unmeaningful terminology is that the other consultants on the Assessment team may also be confused, to the detriment of the Environmental Assessment; and the further risk of confusing the local planning authority or the Inquiry Inspector should also be kept in mind. If the simple questions, why, where, who, what, when and how are seen to be answered clearly, then the summary will achieve its aim.

Why is the development necessary? Economic, social, political and other reasons may be listed under this heading.

Where is the development? This includes the reasons for selecting this particular site and its relationship to the infrastructure.

Who will be affected by the development? This covers the effects on the local communities, the labour market, and the population structure.

What is the development intended for? This covers the processes and end-products of the development, and its place in the local planning authority's development proposals.

When is the development scheduled to start and how long will it last? This covers the construction and operation periods and the after-use (if any) of the site.

How will the development operate? Most of the environmental effects, such as possible pollution and nuisances, will be included in the answers to this question, and as these answers are those most likely to be challenged by objectors, this part of the Summary should receive very careful consideration.

This may seem to be an oversimplified format for a Summary, for which no statutory format has yet been laid down, but if these six questions are fairly and clearly answered, the Summary should not be open to criticism. It is often the practice to simply extract chunks from the team's reports and conclusions in order to form the Summary, and even this may be left until the last moment, producing an unsatisfactory document which does no credit to the team. Since the Summary will be circulated much more widely than the main documents, and is likely to be used by many officials and the Press as a quick reference it is essential that it is written to a high standard of English. A good Summary is written quite separately from the main documents, with a more general audience in mind. As sound informative journalism comes nearest to the ideal standard for a Summary, probably a competent journalist may well be the best qualified person to undertake the writing of the text, and there seems to be no ruling against employing well-known writers for the purpose. Illustrations should be clear and unambiguous and drawn specially for the Summary, not mechanically reduced versions of the main submission drawings, and should be reproducible in black-and-white for the Press.

Explanation or amplification

Sections 3(a), (b), and (c) of Schedule 3 are not discussed here, since the items listed are amplifications of statements made under Section 2. They are fairly closely related to Sections 2(a), (b) and (c) respectively.

Alternatives 3(d)

This comprises the main alternatives studied by the applicant. There are three categories of alternative proposals:

1 alternative sites for the development;
2 alternative processes or construction;
3 alternative methods of dealing with environmental effects.

Alternative sites

Any feasible alternative will have already been examined before the chosen site was subjected to Environmental Assessment, but the reasons for the choice must be described, at least in outline. In the case of industrial projects, the land values, infrastructure availability, level of grant aid, proximity of sources and markets, and the labour supply will be the governing factors in the choice of site rather than environmental considerations, and it is likely that the developer will have to show that no other site is economically viable if he is to overcome environmental objections. In the case of civil engineering projects, it seems obvious to say that only one alignment is the most satisfactory in engineering terms, but the history of successful alternative road alignments put forward by objectors indicates that reliance on engineering factors alone to establish the need for one particular location is not enough, and that *all* environmental factors must be considered.

Alternative processes or construction

The possibility of putting forward acceptable alternative processes is not very great, since the development is probably based entirely on the selected process, but there are alternative forms of construction which may have to be considered. Undergrounding large parts of the development, or using low-level buildings may be suggested by objectors, and the developer should make sure that his architect has fully explored these possibilities before committing the project to one design. Alternative access routes should be considered if possible, since these are often the most contentious part of the proposal, especially if the development is in a built-up area, and the excuse of additional costs for longer access roads may not be accepted by the local planning authority.

114

Alternative methods of dealing with environmental effects

These will have already been well covered in the section on mitigation of adverse effects, and the rejected alternatives need only be briefly described here. Again, any methods of managing effects which are selected only because they cost less than the alternatives are likely to be challenged, and realistic environmental reasons for the methods chosen must be given.

Use of natural resources

These will already have been dealt with in general terms in Section 2(b), so that this section may only need to amplify the exact use to be made of scarce or non-renewable natural resources such as peat, gravel, shingle, mineral springs, or fossil fuels. The time-scale of consumption is important, and any prediction of the amount available of a resource over a given period of time may well be challenged by opposition experts. Strong public feeling is aroused by the use of irreplaceable natural resources so that the developer should restrict his use of these materials to the absolute minimum. One of the Channel Tunnel proposals was severely criticized for the proposed use of Dungeness shingle for construction; Dungeness is an internationally protected shingle beach, and such a proposal is enough to condemn any project immediately.

The emission of pollutants, creation of nuisances, and disposal of waste will already have been covered in Section 2(d).

Forecasting methods and difficulties in compiling information 3(f) and (g)

This comprises forecasting methods used to assess any effects on the environment described in 3(e) and any difficulties encountered in compiling specified information.

The forecasting techniques used will vary from project to project, but for all of them it is important to state, first, the known reliability of the method as demonstrated in previous fieldwork (not laboratory tests) and, second, the level of accuracy to be expected from it. Proven statistical analysis methods of checking data for probability and reliability should be used wherever the type of data allows, and published sources of information should always be verified, since inexperienced researchers are not always as careful about their statistical techniques as they should be. The level of accuracy should always be given even if the range is wider than is desirable, so that work cannot be criticized for pretending to be more accurate than it actually is. For example, if it is only possible to survey a rather small percentage of a population it is better to say so rather than to be criticized later for making judgements based on inadequate evidence. Information may be inadequate for several reasons:

- lack of expert advice, perhaps due to shortage of time or the unavailability of a particular consultant;
- unobtainable for security reasons or because of commercial confidentiality;
- the data may be there but in such a form that it cannot be used without more research – this is often the case with effects such as indirect pollution where work has been done on one aspect but is insufficient to enable the team to draw reliable conclusions for a projection of long-term effects;
- the subject is still at the threshold of knowledge, such as genetic radiation damage or lifelong ingestion of supposedly non-toxic chemicals;
- the existing conditions are simply not known, such as the extent of fossil fuel pockets or the response of fragile eco-systems to apparently non-traumatic changes in the environment.

A certain amount of tact is necessary in describing the reasons why information is inadequate, especially where the source is the local planning authority or a government department, but it should be possible to describe the deficiency without blaming individuals or organizations who may not have been able to provide the exact information required by the Assessment team. In some cases there may be political or administrative reasons for refusing information, and these should also be recorded without allocating individual responsibility.

Section 4 Summary

Section 4 deals with the supplementary non-technical summary related to Section 3 information, and the same comments apply to this Summary as to the main Summary.

116

This chapter is only a brief introduction to the range of subjects which may be included in an Environmental Statement, but it may serve as a starting-point for the Assessment team or individual consultant responsible for its preparation.

Chapter 7

QUALITATIVE ASSESSMENT TECHNIQUES

There is no great difficulty in measuring physical environmental sectors with data which can be analysed to produce reasonably reliable results. Each profession has its own technical methods of measuring physical attributes, ranging from seismic shots for oil exploration to counting migrant birds, and their description is beyond the scope of this book. Qualitative measurements of such sectors as visual effects of a development, loss of amenities, traffic hazards, and job opportunities are far harder to record and to analyse, and this section describes some of the techniques which may be adapted for dealing with these intangible but critical environmental effects.

It is comparatively easy to estimate the increase in traffic flow along a road – although the state of UK motorways does not seem to support this view – but very much more difficult to estimate the increase in the hazards of crossing that road; an increase which depends more on the population mix and age groups than on the number of lorries. It is interesting to note that children brought up in Cumbernauld, which is designed with a high level of traffic–pedestrian segregation, are far more at risk when they go to other towns than children brought up to cope with traffic, and factors of this kind must be included in any analysis. Another example of a failure to analyse quality as well as quantity is the provision of jobs which may be quite unsuitable to the local population: in one town with a shortfall of women's jobs the planners suggested that a local fish-packing industry could usefully be developed – unfortunately the unemployed women turned out to be the wives of senior government scientists and, not surprisingly, they showed little enthusiasm for packing wet fish at six a.m. Similar failures have occurred when government

118

has tried to deploy heavy industry workers into light electronic or clerical jobs.

Probably the most controversial qualitative effects are the visual intrusion of new development into unspoilt landscape, and local feeling will always be very intense on this subject. There is no reason why local people should not be involved in the assessment of such effects if simple and easily comprehended techniques are used, and the results can usefully be compared with the results obtained by consultants. One of the dangers of environmental surveys is that the measurable factors tend to dominate the unmeasurable ones; an infra-red aerial survey will reveal the health of trees and vegetation, but not their role in softening the impact of the development, which might well be the more important factor. Nevertheless, the factual survey will usually weigh more heavily with an assessor than the subjective analysis.

The *Manual for the Assessment of Major Development Proposals* gives a fuller description of some of these techniques, and includes descriptions of many others which may be useful to the assessor. The techniques described here are intended to show the main types of non-quantitative surveys; the variations on these will be more or less suitable according to the type of development to be assessed.

Visual effects

In an Environmental Assessment, one of the best publicized adverse effects is the visual intrusion of the proposed development into the existing landscape. This is most obvious where the development is to take place in open countryside, a coastal area, or close to preserved land such as a Green Belt, Heritage Coast, or National Park. The general public and the environmental amenity groups tend to concentrate on this effect, since it is easily understood and the cost of preparing objections is much lower than the cost of opposing technical effects such as traffic increases and employment losses. Other sensitive places where visual effects are significant are conservation areas, small-scale rural communities, and communities which depend on tourist trade for their livelihood. The scale and intensity of opposition, and therefore the demand for thorough and incontrovertible visual effect assessment, will depend very much on the way the local community see their environment; this is governed by factors such as social class, local employment opportunities, tradition, strength of community organisation, and on the money available for preparing opposition evidence.

There are a number of techniques for assessing visual effects, ranging from computerized analysis of landscape factors and their weighting in the whole landscape, to simple photo-montages showing the landscape before and after the proposed development. It seems only fair to the public (and less costly to the developer) to present visual effect assessment in a form which can be checked by interested parties without great expense, and the more academic techniques should only be used to back up simple methods where the visual effect is very controversial and expert opposition will be encountered. Full descriptions of the sophisticated techniques can be found in professional journals such as *Landscape Research*, *The Planner* and *Landscape Design*.

Zone of visual influence (Fig 7.1)

This technique was developed by Clifford Tandy in the 1970s and can be carried out manually (and slowly) or more rapidly by one of the many computer programs developed for the purpose. The technique is used mainly to show how large an area will be affected by the development, and to simulate the effects of screen planting, earth mounding, and changes of design. It involves projecting the height of the proposed development across a contour map of the surrounding area in order to show which parts will have a view of the development and which will be screened by the landform or by existing woodland and buildings. Manually, this is done by taking radial sections at 5 or 10 degree intervals right round the project and constructing the land profile, with the highest point of the development projected along it to show the exposed and screened areas; the computer can better this by covering every section to give a more detailed analysis. To speed up the manual method, a simple scale can be constructed to mark off exposed areas along the contour lines, thus saving the chore of drawing sections.

Landscape visual quality

What makes a beautiful landscape? Is it the scale, the vegetation, the topography, the presence of water, or a compound of all these? Many efforts have been made to assess the quality of the landscape in order to decide how significant the effect of the project may be; these range from rigid statistical models to 'yes

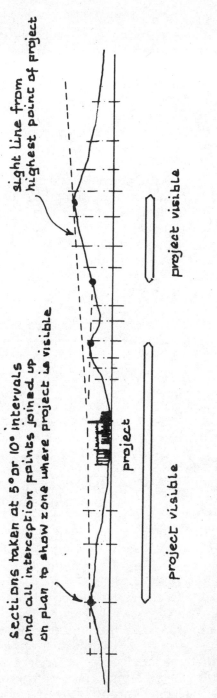

sections taken at 5° or 10° intervals
and all interception points joined up
on plan to show zone where project is visible

sight line from
highest point of project

project

project visible

Project visible

Figure 7.1 Zone of visual influence

121

I like it – no I don't' opinion surveys.

At one extreme is the technique of listing each feature in the landscape; trees, scrub, water, hedges, houses and so on, and then giving each one a value according to a predetermined scale based on its contribution to an ideal landscape. Then the numbers of each element in the area affected by the project are counted, multiplied by their given value, and the results indicate the quality of the landscape in each part of the development area. The advantage of the method is that of providing a standard which can be used to compare landscapes in different areas; the snag is that this method takes no account of the landscape composition, and depends on the originators' opinion of the value of landscape elements. It is also not necessarily true that ten km of hedges make a better landscape than two km, or that half a dozen specimen trees make a worse landscape than twenty.

Next in line is the technique of dividing the landscape into kilometre squares (conveniently done by the Ordnance Survey) and assessing the quality of the landscape in each one against a standard quality determined at the beginning. (Fig. 7.2). The advantage is that the landscape is examined as an entity, and expert landscape designers have no difficulty in holding an ideal in their minds as a comparison. The objection to this method is that the landscape does not change at the kilometre square boundary, and also that the method of scoring the squares implies that certain types of landscape are intrinsically more desirable than others. This is very arguable; landscape designers brought up in Devon cannot see any beauty in the East Anglian scene, and Highland landscapers find the small-scale Home Counties scenery very insipid.

Some landscape designers consider that landscape is a purely aesthetic experience which cannot be subjected to arithmetical calculation, and can only be analysed in terms of its impression on the viewer. One method developed by Derek Lovejoy and Partners is based on an analysis of the landscape as seen in views from a selection of key points around the development, the scoring being based on the viewers' opinion of the quality of individual landscape elements, combined with an overall appreciation of the landscape. (Fig 7.3). This method is subjective, and depends on an honest and skilful appraisal by a number of landscape designers whose scores are averaged to give the final result. The advantage is that the landscape 'as seen' by the public is studied, so that artificial weightings are not introduced into the analysis, and also that interested groups can repeat the exercise for themselves; the

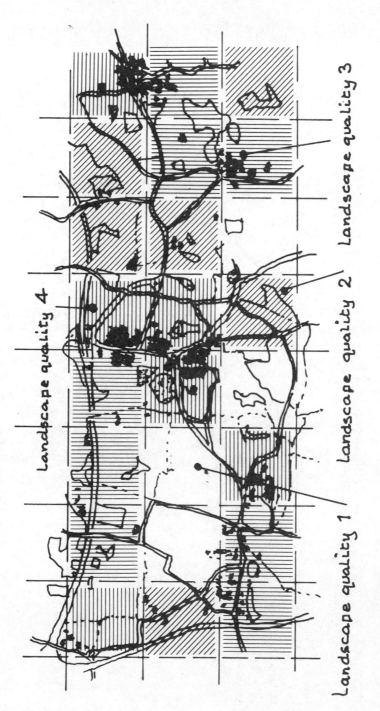

Landscape quality 4   Landscape quality 2   Landscape quality 3

Landscape quality 1

Figure 7.2 Kilometre squares

123

124

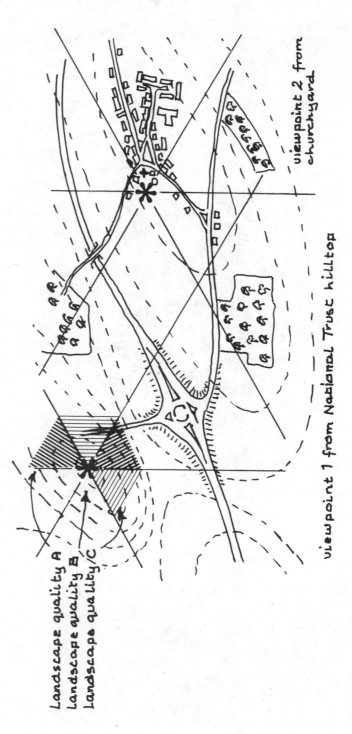

Landscape quality A
Landscape quality B
Landscape quality C

viewpoint 1 from National Trust hilltop

viewpoint 2 from churchyard

Figure 7.3 Viewpoints

disadvantage is that, in practice, only a limited number of view-points can be taken, and therefore landscape concealed by the landform is not assessed.

Balancing the values of effects

After all the individual assessments have been made, the work of collating and cross-checking the results has to be done. This must necessarily be the job of one person, or no more than two or three, who can grasp all the main points of each report and bring them into a coherent Environmental Statement. Each effect has to be compared with other effects and the level of significance of each must be assessed. Despite the resident statistician's firmly held conviction that everything can be turned into figures, this really cannot be done mathematically (ask him if the three most important factors in a man's life – his career, his wife, and his friends – have ever been decided by statistical calculations). The greatest danger in Environmental Assessment is the artificial weighting of results to produce a seemingly valid conclusion, and any weighting applied to raw data must be genuinely derived from that data and not imposed subjectively.

Sieve maps

This technique consists of preparing maps of the area, each showing a different sector of the environment; water bodies, housing, woods, farmland, infrastructure, protected areas, and so on. These maps are then laid over one another, either manually or by computer graphics, to show which areas are least vulnerable to intrusion by development. (Fig. 7.4) It is unlikely, in a highly developed country like the UK, that *any* area can be found which is completely free of any constraint on development, but the technique is very valuable in showing the least heavily loaded areas and, more importantly, for showing why alternative sites have not been selected. Should the sieve maps show an apparently unloaded site, further maps showing economic travel distances or land values may account for the lack of previous development.

Matrices

These are used to show the interrelationships between environ-mental sectors and the development – known to the Regulations as direct and indirect, or secondary effects. Matrices have not

125

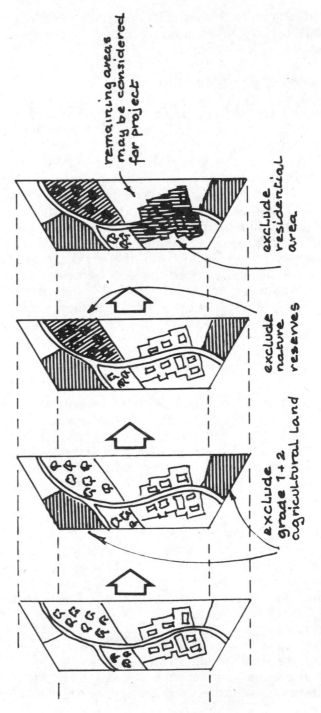

Figure 7.4 Sieve maps

been developed very far as a tool of Environmental Assessment, partly because they are time-consuming to construct, and partly because a really good matrix will show many indirect effects which are not really the responsibility of the developer. The simplest form of matrix merely indicates where a relationship between the project and environmental sectors occurs, and provides a check on the Assessment team's work, though without giving any information as to the nature of the relationship. It has a useful though limited function. (Fig 7.5).

Environmental sectors	Environmental effects			
	Pollution	Land take	Visual effect	Severance
human beings	*	*	*	*
flora	*			
fauna	*	*		
soil				
water	*	*		
air	*			
climate				
landscape		*	*	
material assets		*		*
heritage		*	*	*

Figure 7.5 Key matrix

Far more profitable, though more labour-intensive, is the complex heuristic matrix which can display every direct and indirect relationship of the development with its environment. This can be used to simulate changes in the scale and function of the development and is thus useful as a tool in making decisions on the project at an early stage. The technique is basically simple to use, involving no complex manipulation of data but, properly carried out, it forms a model of the effects and side-effects of the development and their relationship to each other. This is a simplified version of the more sophisticated Ecoplanner method developed at Heriot Watt University for modelling and analysing urban and regional systems. In this simpler version, the project itself, and each of the environmental sectors to be considered, occupy a square on the diagonal of a matrix. The effects of the project on each sector are then entered in the square where their coordinates meet, reading clockwise in all cases. (Fig. 7.6).

Likewise, the secondary effect of one sector on the project or

128

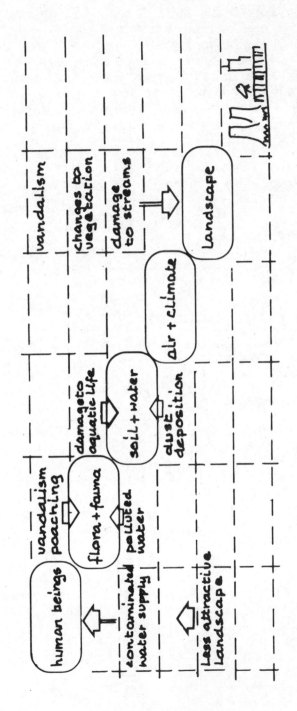

Figure 7.6 Effects matrix

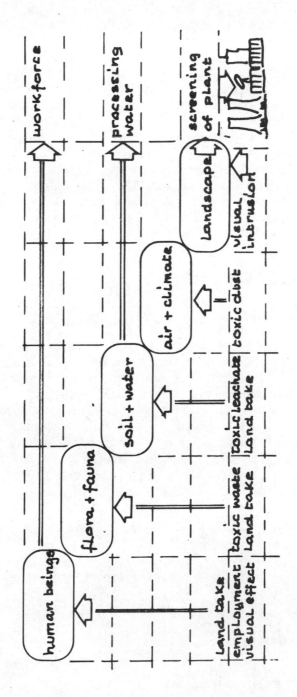

Figure 7.7   Secondary effects

129

on the other can be entered on their coordinates and read off. In theory, the matrix is infinite in scale, covering local environmental sectors close to the project and extending further to regional and national sectors; in practice it is enough to consider the local significant effects only. (Fig. 7.7).

The advantage of this method is that *all* effects can be recorded, usually by means of key references on the matrix to a database, and that each member of the Environmental Assessment team can enter his own data and read off the other members' results. The consultant responsible for the Environmental Statement can then assimilate all the information about each sector without difficulty. Like all other models of real life, this matrix is only as good as the people who make it but, well handled, it can ensure that no awkward questions are left to emerge unexpectedly at public hearings or planning inquiries.

Data collection

Opinion surveys

It is not easy to ensure that a questionnaire on some aspects of the development is fairly and comprehensibly designed, and expert help in the design of questions and of the form itself is advisable. Forms for computer analysis need to designed with the chosen software in mind. The best way to start is to decide exactly what it is that the team wish to ascertain, and to work back to the questions from there. It may be possible to cover several environmental factors in one questionnaire, but most people will only fill in one side of a sheet, and more extensive replies may need

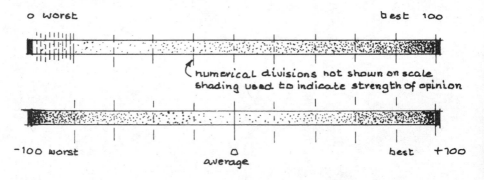

Figure 7.8 Scoring scales

the inducement of a small gift in the form of pens or plastic goodies of the sort familiar in all consultants' Christmas mail. Very few surveys of qualitative sectors can be dealt with by straight 'yes or no' questions and answers. There are two useful methods for dealing with imprecise answers: using ranges, and using semantic differentials. The range method offers the respondent a scale on which to mark his evaluation of the item; usually a scale of 100 points appears to offer a better choice than a scale of 10 points. The ends of the scale may be figures or words according to the subject. (Fig. 7.8).

Semantic differentials offer the respondent a choice of words to describe each item, ranging from negative or derogatory words at one end of the scale to positive or commendatory words at the other. These words should be carefully selected to give the exact shades of meaning required by the team; 'better', 'bad', 'worse' are not at all helpful. The words need not be technical, since the English vocabulary in common use contains enough words to cover most opinions; in some areas where there is a

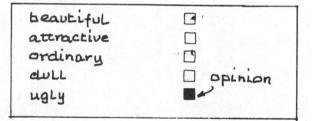

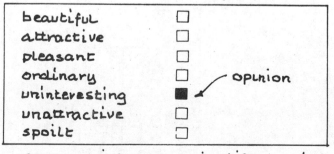

Figure 7.9 Semantic differentials

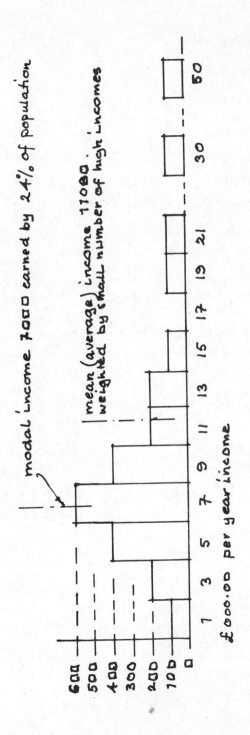

Figure 7.10 Modal scores

significant linguistic group, it may be necessary to translate the questionnaire into Urdu, Welsh, or Polish, in order to be fair to all respondents. (Fig. 7.9).

Averages

The average man, the average reply, and the average age all come into qualitative assessment analysis. Usually this is taken as the simple arithmetical mean obtained by adding the figures and dividing by the number of units; add up the incomes of 2,500 people, divide by 2,500, and it appears that the population have an average income of £x per annum. This simple calculation is often the only one seen in assessments and, even where statisticians are not employed, there are other useful ways of calculating central tendencies. Besides the arithmetical average, there is the modal score, which is the income earned by the largest number of people out of the 2,500; it might be £10,500, which suggests that a good proportion of the population is adequately paid. On the other hand, if the modal income for the 2,500 people is only £6,000, this suggests that the population is not so well off. (Fig. 7.10).

If these figures are plotted as a curve, they will usually approximate to the 'normal curve' which is a form found in the majority of surveys, and shows a normal distribution of data. In a normal curve 68 per cent of the population will lie in the middle third. Some surveys reveal 'skewed curves' showing a marked bias towards one end of the scale or the other. If the population is unbalanced towards riches or poverty, the curve would be 'skewed', showing the percentage of people who fall towards one extreme or the other. Again, this is very useful in analysing opinion polls, since it shows how far the population leans towards either of the extremes set by the questioner. (Figs. 7.11 and 7.12).

A variant on this graph is the percentile graph, which shows the percentages of the population in each income bracket instead of the actual numbers. If divisions of 1 per cent are used, the graph is calculated in 'percentiles'; if divisions of 25 per cent are used, the graph is calculated in 'quartiles'. (Fig. 7.13).

These simple calculations are good enough for rough working to estimate the significance of an effect, but more sophisticated statistical analysis, which covers factors such as bias, reliability, and small samples, is needed for decision-making on critical issues, and expert statistical help should be employed. One

134

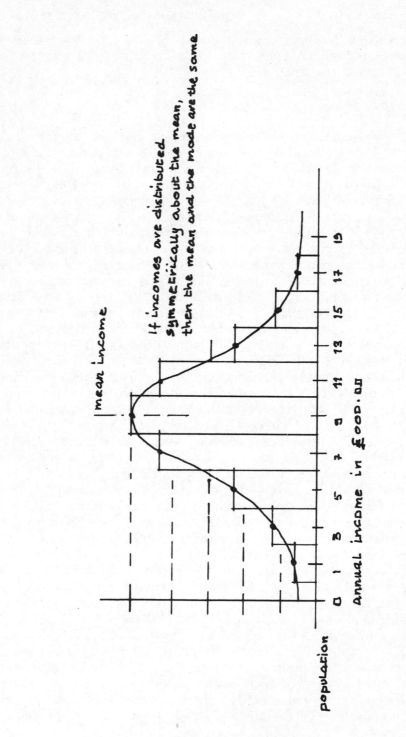

Figure 7.11 Normal curve

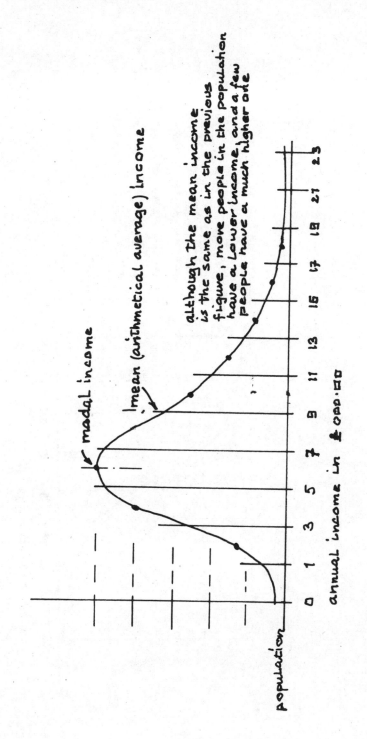

modal income

mean (arithmetical average) income

although the mean income
is the same as in the previous
figure, more people in the population
have a lower income, and a few
people have a much higher one

population

annual income in £ooo.ɒɒ

Figure 7.12 Skewed curves

135

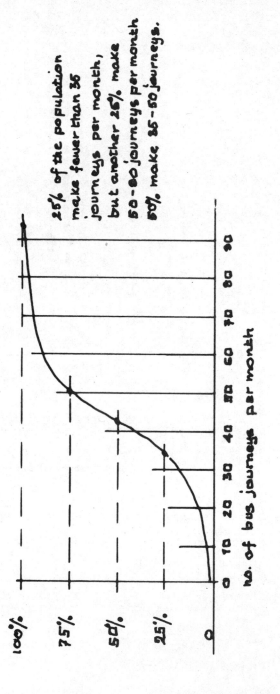

25% of the population make fewer than 35 journeys per month, but another 25% make 50-80 journeys per month, 50% make 35-50 journeys.

no. of bus journeys per month

Figure 7.13   Percentiles

136

survey of medical research reports showed that 32 out of 62 reports had statistical errors, 18 of which were serious; and it seems obvious that an Environmental Assessment containing such errors could be challenged in an inquiry or court.

In many Assessments, it is necessary to determine whether or not there is a correlation between two or more factors, such as distance from public transport and higher levels of car ownership, or between proximity to noisy roads and school performance. A description of standard errors, regression analysis and probability theory is beyond the scope of this book, but may be found in any elementary book on statistics. There are many pitfalls in calculating correlations between factors, not least that of producing a correlation between irrelevant factors which has a false appearance of validity. For example, a close correlation between children going to school by bus and bad behaviour in school was taken to show that buses caused bad behaviour – an entirely illogical conclusion. If correlation checks are made, *all* factors should be checked against each other, not only the two considered to be relevant by the assessor. The correlation of factors in an Environmental Assessment may be a very important part of the Assessment, since assumptions on the relationship between pollution and ill-health, or goods traffic and accident rates, need to be tested before any decisions are taken about possible mitigating measures.

Statutory information

Although there is provision in the Regulations for charges to be made for information supplied by official bodies to the Environmental Assessment team, the people who actually prepare the replies have no inducement to extend their already busy work schedule and, at the least, a courteous and appreciative letter should be sent to them. It is helpful and more efficient to set out the request for information in an organized form which covers the critical data needed for the Assessment, rather than the 'write down all you know' type of enquiry, though further information should always be invited. Even if the reply is in the form of a massive document covering all aspects of the organization's work, the useful information should be extracted for the benefit of the team, with the document references given as appropriate. Since there are a very large number of bodies who must be consulted, all requests should have the same format as far as possible, so that information can be easily

137

seen and collated. A useful set of headings, not necessarily in this order, might be:

- name of environmental body and name and phone number of contact;
- name of team member responsible for collecting information;
- environmental responsibilities of body (conservation, pollution control, water management);
- statutory environmental powers of body as far as the development is concerned;
- advisory environmental powers of body as far as the development is concerned;
- environmental effects for which the information is required (these may not be known at the time of asking, but the type of effect – flora, soil, water, human beings – can be specified);
- date and original sources of information held by the body (committee decision, research paper, opinion poll, data extrapolated from national data bank);
- type of information: 'hard' (reliable provable statistics), 'intermediate', (reliable but not strictly provable), or 'soft' (opinions or subjective judgements);
- future policies or programmes of work and research relevant to the development;
- permission to quote from source or not;
- informal comments from body which may not be treated as formal information;
- if the information is held in a computer database, the disk and file reference;
- if the information is on drawings or photographs, a summary of the subject and the file reference to the negatives and copies;
- references to other sources of information.

Key people

This section is more appropriate to Environmental Assessments for smaller developments, or those in self-contained communities, than to the 'Big One' or long-distance roads or railways which pass through many different areas. In every community, trade, or field of expertise, there is always one, sometimes two, people who hold the key to the information on a particular environmental sector. Informal discussion with local planning authority officers, the public library, or the local newspaper, will lead the Assessment team to these people, and they may be able to supply unpublished or unknown data which will help to determine the effects of the development. Two examples may make this clearer.

1 An area of land along a river had to be checked for the presence of protected flora and fauna before flood relief works could be proposed. This would have involved surveys carried out over several seasons, but the library produced a local university postgraduate student who had carried out a comprehensive survey for his thesis, thus saving the Assessment team time and money, whilst providing a fuller survey than would have been possible within the assessment program.
2 A housing development adjoining a railway in the Green Belt was intended to provide multi-storey flats of conventional construction on apparently normal ground. The local grapevine produced a railway historian who knew that the site had a very peculiar unstable geology which had caused severe problems during the building of the railway. The proposed housing construction would have resulted in a major structural failure.

This unrecorded information is not usually critical to the development, but it may reinforce or contradict the more formal data accumulated by the team, and it is unwise to ignore it altogether. Some environmental sectors which may yield information in this way are:

- ethnic or cultural biases which may affect the public attitude to the project and which are not obvious to the team – for example, a project relying on a female workforce may have staffing difficulties in a predominantly Muslim area;
- historic ownership or rights which are not recorded in public archives, such as the King's Cross development land ownership;
- very local geological, hydrological, or meteorological factors which do not appear in national records, such as local fogs, floods, or frost pockets;
- rare flora or fauna which must be conserved;
- archaeological sites which have not been officially excavated or recorded, but which may be known to local societies or historians – many developers have run into difficulties in historic cities when work has started before the sites have been checked;
- local customs and organizations which operate in an

139

unorthodox way, and traditional occupations which make it difficult to recruit a satisfactory workforce. This is particularly true in island communities and well integrated ethnic groups.

These are only a few of the non-quantitative ways of obtaining Environmental Information; many more will be developed by the Assessment team as it gains experience, and many more will be published as the art of Environmental Assessment comes of age.

Chapter 8

CONCLUSION

The practitioners of Environmental Assessment, both consultants and authorities, are still exploring the principles, contradictions, and uncertainties of the way forward to good environmental analysis. There are still many problems.

1 The clash between national and local environmental issues. Does national economic advantage cancel the damage to a local environment? Do national environmental protection policies override the loss of local jobs? An Environmental Assessment only deals with one project in one place, and a method of dealing with the conflict between national and local policies has yet to be designed. Sometimes it is advisable to go right back to the original decisions on the development and to reconsider first principles; a course which most clients are most unwilling to consider, but which may be the most profitable in the end.
2 The difficulties of judging which environmental impacts of the development are likely to be significant and which are likely to be insignificant; for whom, for how long, and to what degree. How far can accurate projections of impacts be made? How justifiable is the cultural, economic or social cost of the significant impacts of a development on a natural or human community?
3 The balancing of quite diverse factors such as employment, visual impact, and use of natural resources. Everyone sets different values on their environment, their job, and their relationship with the natural world; someone has to decide which values shall take priority.

4 The problem of providing enough expertise and money to check an Environmental Assessment, and to monitor the development during construction and operation. Should the local planning authority, financed by the local community, carry this load, or should the cost be borne by the developer? Either choice seems unfair, yet someone must pay for proper control of the environment, or else the whole process becomes a pointless exercise in bureaucracy.

5 The decision on the limits of an Environmental Assessment. Should it cover just the surrounding area, or should the impacts be traced as far as they appear to be significant? A project may be small in size, but its repercussions on rail links the other side of the country, the dumping of wastes in distant valleys, or the pollution of water many miles downstream are still significant to some other part of the ecosystem.

6 What are the thresholds of tolerance? Is it for the developer to prove that some aspect of his project will be harmless, or is it for the controlling authority to decide whether an impact is tolerable or not? Even for those impacts which are measurable, the thresholds may vary in time and place; every camel has his own set of weights and measures for the last straw. One development, two developments, three developments may be absorbed into the local ecosystem, but one more may well tip the whole human and natural system into disarray as the supporting infrastructure of a district, or the natural regeneration system of a river, breaks down from the overload.

7 How are multiple developments, in which some projects merit an Environmental Assessment and others do not, to be controlled? Should the entire complex be subjected to assessment, possibly causing a lot of unnecessary work, or should each part be treated on its individual merit, thus denying the relationship between one part of the development and the others?

8 How far are the promoters of one development to be held responsible for adverse effects generated by later developments consequent on their own, which they could have foreseen but not controlled, and should they be held liable for changes of heart by other bodies working with them on the assessment? A case in point is the proposed rail link from the Channel Tunnel to London, where British Rail originally said nothing about new lines, nor indeed about high speed-trains. The decision to introduce new high-speed tracks to London, and to increase the number of freight trains, appears to have been taken *after* the Channel Tunnel proposals were accepted, despite

the close relationship between cross-channel and UK traffic. Even the decision that all freight and passenger traffic should go to London seems to be a later idea, since this concept does not form part of the Channel Tunnel Environmental Statement. Had an Environmental Statement for the associated rail links been integrated with the Tunnel Environmental Statements, both projects might possibly have been modified. Perhaps this type of multiple development is a situation where the Secretary of State could usefully exercise his powers to compel the relevant bodies to prepare interrelated Environmental Statements.

The answers to these questions will be found – slowly perhaps, and probably not without some bitter argument and some revisions to the existing legislation – and eventually a body of experience and accepted procedures will be created. Future consultants in the field of Environmental Assessment will probably wonder why we found such difficulties and hardships in the preparation of simple Environmental Statements, but until then this book may help to answer some of the queries which arise in the minds of those who, one way or another, have to deal with the problems of Environmental Assessment.

Index

Italic numerals refer to the figures

Index

Index